D0641056

WIND POWER IN VIEW

ACADEMIC PRESS
SUSTAINABLE WORLD
SERIES

SERIES EDITOR
RICHARD C. DORF

University of California, Davis

The *Sustainable World* series concentrates on books that deal with the physical and biological basis of the world economy and the dependence on nature and the tools, devices, and systems used to control, develop, and exploit nature. Engineering, as the developer and the user of these technologies, is the key element in a process of planning for a sustainable world economy. If the industrialization of the world is to continue as a positive force, the creation and application of enviromentally friendly technologies should be one of the highest priorities for technological innovation in the present and the future.

This series will include titles on all aspects of the technology, planning, economics, and social impact of sustainable technologies. Please contact the editor or the publisher if you are interested in more information on the titles in this new series or if you are interested in contributing to the series.

Current published title:
Technology, Humans and Society: Towards a Sustainable World, edited by Richard C. Dorf, 500 pages, published in 2001.

WIND POWER IN VIEW

ENERGY LANDSCAPES
IN A CROWDED WORLD

EDITED BY

MARTIN J. PASQUALETTI
PAUL GIPE
ROBERT W. RIGHTER

ACADEMIC
PRESS

An Elsevier Science Imprint

San Diego   San Francisco   New York   Boston
London   Sydney   Tokyo

This book is printed on acid-free paper.

Copyright © 2002 by Academic Press

All rights reserved.
No part of this publication may be reproduced or transmitted in any form or by
any means, electronic or mechanical, including photocopy, recording, or any
information storage and retrieval system, without permission in writing from the
publisher.

Requests for permission to make copies of any part of the work should be mailed
to the following address: Permissions Department, Harcourt, Inc., 6277 Sea
Harbor Drive, Orlando, Florida 32887-6777.

Explicit permission from Academic Press is not required to reproduce a
maximum of two figures or tables from an Academic Press chapter in another
scientific or research publication provided that the material has not been credited
to another source and that full credit to the Academic Press chapter is given.

Academic Press
A Reed Elsevier Imprint
525 B Street, Suite 1900, San Diego, California 92101-4495, USA
http://www.academicpress.com

Academic Press
Harcourt Place, 32 Jamestown Road, London NW1 7BY, UK
http://www.academicpress.com

Library of Congress Catalogue Card Number: 2001096353

International Standard Book Number: 0-12-546334-0

Printed in China

02 03 04 05 06 07 RDC 9 8 7 6 5 4 3 2 1

CONTENTS

PART II

WIND POWER ON THE LAND

1

2

3

THE WIND IN ONE'S SAILS: A PHILOSOPHY   59

GORDON G. BRITTAN, JR.

PART III

WIND POWER IN NORTHERN EUROPE

4

WIND LANDSCAPES IN THE GERMAN MILIEU   83

MARTIN HOPPE-KILPPER AND URTA STEINHÄUSER

5

SOCIETY AND WIND POWER IN SWEDEN   101
KARIN HAMMARLUND

6

A FORMULA FOR SUCCESS IN DENMARK   115
FRODE BIRK NIELSEN

7

LANDSCAPE AND POLICY IN THE NORTH SEA MARSHES   133
CHRISTOPH SCHWAHN

PART IV

WORKING WITH THE WIND

8

LIVING WITH WIND POWER IN A HOSTILE LANDSCAPE   153

MARTIN J. PASQUALETTI

9

DESIGN AS IF PEOPLE MATTER: AESTHETIC GUIDELINES FOR A WIND POWER FUTURE   173

PAUL GIPE

PART V

AFTERWORD

1O

ACKNOWLEDGMENTS

We are grateful to the Rockefeller Foundation for providing the venue and on-site support for the workshop where the original papers were read and discussed, and to the United States National Renewable Energy Laboratory for providing travel funds. Thanks are also due to Joan Pate, without whom this book would have never been possible. Her interest in the relationships between rising population and declining environmental quality provided much of the initial impetus for the development of this book.

MJP
PG
RWR

I

INTRODUCTION

O

A LANDSCAPE OF POWER

MARTIN J. PASQUALETTI, PAUL GIPE,
AND ROBERT W. RIGHTER

To the pundit who said "there is no such thing as bad publicity," we offer wind power as an exception to the rule. Although it is now blossoming into the fastest-growing energy resource in the world, wind has also been labeled a competitor. Despite its several attributes, it has been dogged by the criticism that it interferes with aesthetic values, that it changes the surroundings too much for comfort, and that it transforms natural landscapes into landscapes of power.

Such a reservation should not be suprising, for it is at the center of the perennial question of how to live in greater balance with our environment. To what degree are we willing to give up landscape quality for qualities of life? Do we want forests or firewood? Green hills or black coal? Rivers to admire or dams to provide us the electricity to run our cities? Is there a way to blend these two needs? We are not asking anything new; rather, it is a question of how to best balance the nature we want with the energy we need.

Although this dilemma is not new, we are facing it more frequently because populations are growing and the amount of open land is shrinking. Is there room enough to meet both needs, or will we have to choose? Those with the low standard of living common in most parts of the world always favor energy supply, but in the United States and Europe people are prosperous enough to be genuinely stymied by the choice they face. Ironically, in this newest version of an old choice, we are focusing not on a

Copyright © 2002 by Academic Press.
All rights of reproduction in any form reserved.

fuel such as coal with a dirty reputation, but on an alternative energy resource with a benign image. It is renewable and releases no pollutants; it can be installed in small, affordable increments; and the potential contribution it can make in industrialized and developing countries is impressive. What is there not to like? The answer to that question is simple: wind turbines are unavoidably visible, even intrusive. They interfere, some argue, with local landscape aesthetics.[1] In the final analysis, despite wind power's many advantages, its potential contribution, and its prospects for rapid growth, by the mid-1990s it had become obvious that its "landscape problem" was here to stay. The time was right for a focused discussion of wind energy landscapes.

In an effort to facilitate this discussion, the Rockefeller Foundation made available its conference center on Lake Como, Villa Serbelloni, for a 10-day period of intense dialog among an international group of wind power experts from several disciplines, including geography, engineering, landscape architecture, history, industrial design, the visual arts, and philosophy. The villa itself overlooks the picturesque, northern-Italian village of Bellagio, a popular holiday locus since the time of Pliny the Younger. Its grounds include an Italianate garden of ordered olive trees and red-tiled stone buildings and are bordered to the north by a dark forest, replete with hidden grottoes. From the villa are views of the gardens and steep-walled valleys of Lago di Como and Lago di Lecco. The only blemish on this bucolic scene is the urban pollution that sometimes wafts northward from the Po River Valley and blots out the sparkling lakes. The juxtaposition of visible industrial waste and the crisp natural beauty at Bellagio made it an ironically ideal place to consider the edgy relationship between the charisma of landscapes and the costs of technology.

Although wind power has provided motive force for centuries, its large-scale application to generate electricity has occurred only in the past two decades. During that period this use has spread most quickly in Europe and the United States, and understandably the competition for space has as well. For example, staffers of Denmark's largest environmental organization are encouraging the placement of machines out to sea so they "won't be seen."[2] To the southeast, the German Association for Landscape Protection has become increasingly strident in its efforts to shield landscapes from wind power's "depredations."[3] Also in Germany, no-nonsense books about the social costs of wind power are increasingly available, including Otfried Wolfrum's *Wind Energy: An Alternative It Isn't*.[4] To the west, wind power's landscape intrusion has been reported as the most important factor in the opposition it is receiving in the Nether-

FIGURE 0.1   Cartoon illustrating public reaction to a proposal by the
Tennessee Valley Authority to erect wind turbines on Lookout Mountain above
Chattanooga, Tennessee. (Reprinted with permission of the *Knoxville News-
Sentinel* Company.)

lands. Across the Channel, opponents in England have labeled wind
turbines "lavatory brushes in the sky." And in the United States, objections to wind power have included determined opposition in Wisconsin,
legal suits in Palm Springs, angry confrontations north of Los Angeles,
and sardonic cartoons in Tennessee (Figure 0.1).

Part of the increasing attention paid to the environmental impacts of
wind power development is resulting from its quickening pace and
growing contribution. By 2001, wind turbines around the globe were
generating 30 terawatt-hours (TWh) of electricity.[5] About one-fifth of that
was being produced in North America (Figure 0.2). By 2002, worldwide
wind generating capacity was expected to exceed 25,000 megawatts
(MW), with the lion's share installed in Europe (Figure 0.3, Table 0.1).[6]
The European Wind Energy Association hopes to install 40,000 MW by
the year 2010, enough to supply electricity to about 50 million people.
With growth of new installations booming, principally in Denmark,
Germany, and Spain, they will likely meet that target.

One of the most important factors in the accelerated interest in wind
power stems from its growing economic force. More than US$6 billion of
new wind turbines are expected to be installed worldwide in 2001, and

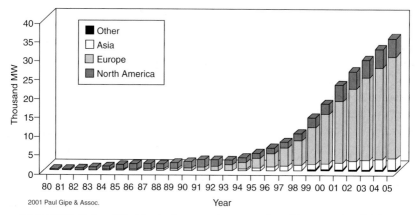

2001 Paul Gipe & Assoc.

FIGURE O.2   World wind generating capacity. (Courtesy Paul Gipe.)

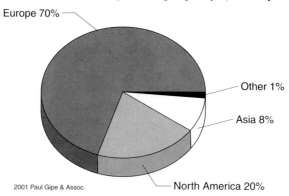

2001 Paul Gipe & Assoc.

FIGURE O.3   World wind generation. (Courtesy Paul Gipe.)

TABLE O.1   World Wind Energy Market Leaders

	Year 2000 capacity additions MW	Year end 2000 total installed capacity MW
Germany	1668	6113
Spain	1024	2821
United States	67	2570
Denmark	603	2341
India	169	1204
Netherlands	40	473
Italy	147	431
United Kingdom	63	424
China	84	309
Sweden	45	267

Source: Paul Gipe & Assoc., BTM Consult.

existing wind turbines are churning out more than US$1 billion in revenues from the sale of electricity annually.

And enthusiasm for wind power continues to grow. The WorldWatch Institute has identified wind power as the world's most attractive renewable energy resource, and Greenpeace has launched a glitzy campaign to address global warming by encouraging the installation of thousands of wind turbines at sea off the coast of northern Europe. Denmark's Energy Minister Svend Auken announced at a global warming summit in Washington, D.C., that his small Scandinavian country would provide 50% of its electricity by 2030 with renewable energy, most of it from wind. And that would mean a substantial increase in the significance of the wind industry to the Danish economy; already the export of Danish wind turbines rivals that of the country's renowned ham. These endorsements are a sign not only that wind power will continue to grow as new countries join the wind fraternity, but also that many people find wind turbines on the landscape acceptable (Figure 0.4).

Considering wind power's recent history, one might naturally be curious about its future. Will its detractors stunt its growth? Is its future in doubt? More to the point, can we do without it? Consider its potential. According

FIGURE 0.4   Jet skier amid wind turbines in the percolation ponds near Palm Springs, California. (Courtesy Martin Pasqualetti.)

to the United States Department of Energy, only 0.6% of the land area in the lower 48 states would be needed to produce 560,000 million kWh per year. Assuming that the typical American household consumes 12,000 kWh per year, this would be enough electricity to supply more than 45 million households. If all Americans used as little electricity as Californians, this would meet the needs of 90 million homes. If Americans used no more electricity than the average European, the same amount of wind-generated electricity could supply nearly 200 million households. There is enough wind resource in North Dakota alone to supply at least one-third of electrical demand of the entire United States.

Given the abundance of the wind resource, the adaptability of wind power to existing land uses, its nonpolluting character, and its increasing cost effectiveness, the wind power industry is bullish about its future. However, the industry, especially in the United States, has been less than successful in convincing the public that wind power can or *should* be used more extensively. One of the critical questions, then, is to identify what must be done if wind power is going to fulfill its potential.

One of the aesthetic problems confronting the new face of wind power is that the turbines are only distant cousins to the familiar windmills of the Netherlands with which many are comfortable. Dutch millwrights used naturally available materials, especially wood and canvas, to make the blades. The rotor blades on modern wind turbines, in contrast, use fiberglass or high-strength wood composites covered with a glossy protective coating. Where traditional windmills were often squat and used timber frames clad in wooden shingles, stone, or brick for towers, today's machines are usually tall, slender columns of steel. Where Dutch windmills could be colorfully painted, wind turbines today are usually found in white or muted shades of gray. Modern wind turbines are, in a word, different (Figure 0.5). And, if we are to generate significant amounts of electricity, they will be plentiful.

The new turbines come in many sizes, from those you can hold in your hands to 2.0 MW giants. The typical 250 kW wind turbine uses a rotor 25 to 30 meters (80 to 100 feet) in diameter and is installed on towers 30 to 40 meters (100 to 130 feet) tall. At the upper end of the spectrum, the rotors on megawatt-size turbines span 60 to 70 meters (200 to 230 feet). As the wind industry begins the new millennium, the most widely used size ranges from 600 to 900 kW. Each blade on these machines is 22 to 25 meters (70 to 80 feet) long. Though most have been installed on towers roughly equivalent to their rotor diameter, some turbines with rotors 50 meters in diameter have been installed on towers 100 meters (330 feet) tall, that is, the tower is the length of a football field.

FIGURE 0.5   Old and new. Modern wind turbines use different designs and different materials than the traditional Dutch windmill in the foreground. Groningen province, the Netherlands. (Courtesy Paul Gipe.)

The extent and specific configuration of modern wind turbines vary with terrain and local planning regulations. In some locations, developers have planted wind turbines in row upon row, using flat landscapes as a farmer might approach a freshly plowed field. These large, often rectangular arrays have given rise to the expression "wind farm." Such arrays can be seen at Gestenge on the west coast of Denmark (Figure 0.6) and on the edge of the Colorado desert near Palm Springs, California (Figure 0.7). In steep terrain, such as in California's Tehachapi Pass southeast of Bakersfield, wind turbines are arrayed in rows along the ridge tops, making them particularly noticeable (Figure 0.8). The placement, the number, and the location of wind turbines have produced controversy as well as electricity.

FIGURE 0.6   Gestenge, a rectangular array of wind turbines on a flat former lake bed in northwestern Denmark. (Courtesy Paul Gipe.)

FIGURE 0.7   Rows of wind turbines looking south toward Mt. San Jacinto and Palm Springs, California, in 1998. (Courtesy Martin Pasqualetti.)

FIGURE O.8   Concentrated ridgetop arrays in Tehachapi Pass. (Courtesy Martin Pasqualetti.)

Because wind development most often occurs in rural areas, it tends to inflame preexisting rural–urban conflicts. In some cases, rural residents resent urban developers who build wind projects in their midst. In other cases, rural residents who want wind turbines for their own use, or for the economic development they promise, resent what seems like meddling by urban residents intent on preserving the countryside for its recreational and scenic value.

Despite the convenience and appropriateness of using rural lands, wind development is not precluded from urban areas. Indeed, many wind turbines in Europe are located within villages and even within large cities. In Denmark there are three cooperatively owned wind power plants within metropolitan Copenhagen (Figure 0.9). In Denmark and the Netherlands, wind turbines are visible near lock gates and busy highways, at fast-food restaurants, in the parking lots of shopping centers, and at parks and playgrounds, as well as offshore and on dikes (Figure 0.10).

Although the development of wind power has never had clear sailing, its rapid expansion in the 1990s is bringing a reluctant industry face to face with an awkward reality: not everyone wants a wind turbine in their backyard, especially when that wind turbine is not their own. One of the contributors to this volume, Robert W. Righter, encountered this phenomenon when researching his book on the history of wind energy in the

FIGURE 0.9   Tourists who photograph Copenhagen's Little Mermaid capture more than expected. In the background, beyond the structures of a working harbor, are wind turbines of the Lynetten cooperative. The 600-kW wind turbines stand on a breakwater within Denmark's capital and are owned by city residents. The turbines are visible from most prominent vantage points within Copenhagen, including Christiansborg, the seat of the *Folketing*, Denmark's parliament. (Courtesy Paul Gipe.)

United States.[7] Despite broad support for renewable energy in general and wind energy in particular, he found many cases where opponents successfully stopped wind energy development in its tracks. What intrigued Righter as a historian of environmental activism was opponents' pronouncement that despite their support for wind energy in principle, various locations were inappropriate. The wind turbines, they believed, simply should always be put "somewhere else" or at least "not in my backyard" (NIMBY) (Figure 0.11).

FIGURE 0.10   Wind power plant in a linear array following a dike north of Urk, Noordoostpolder, the Netherlands. These medium-size wind turbines use a rotor 25 meters (80 feet) in diameter to power 250-kW generators. Dutch tourists for a Sunday morning stroll along the public footpath. (Courtesy Paul Gipe.)

The NIMBY reaction to wind energy that Righter saw so clearly in the United States prompted him to seek the help of geographer Martin J. Pasqualetti and wind energy advocate Paul Gipe in organizing a multi-disciplinary symposium to discuss how this promising technology could be reconciled with the sometimes conflicting demands of nature and need. In response to a proposal by Righter, the Rockefeller Foundation awarded 10 fellowships for a 10-day retreat at its Villa Serbelloni. The accidental symmetry of the "10 for 10" illustrated the topic at hand, finding unity and order in human-altered environments.

Given that the problems being faced by the promoters of wind power are a complicated mix of technology, planning, aesthetics, engineering,

FIGURE O.11   Three medium-size wind turbines installed in a cluster at a small factory on Germany's central plateau (Hoher Westerwald, Hesse). Though clearly in someone's backyard, wind power in Germany has not faced the same opposition as in Great Britain. (Courtesy Paul Gipe.)

and policy, and that this mix existed both in the United States and in Europe, it followed that the best discussion of issues of wind energy compatibility should include a cross-disciplinary, international group who could share the perspectives of their countries, their personal experience, and their research. Righter, Pasqualetti, Gipe, and Montana State University's Gordon Brittan represented the United States. Pasqualetti has been studying the relationships between energy and land use for 25 years at Arizona State University. He conducted one of the earliest surveys of public attitudes toward wind power and brought his knowledge of American public opinion to Bellagio. Gipe writes and lectures about wind power. He contributed his firsthand experience explaining to the public both the problems and the promise of the technology. Brittan may be the only philosophy professor in the world who operates his own wind turbine.

The other participants were European: two Scandinavians, two Germans, and one each from Britain and the Netherlands. Swedish geographer Karin Hammarlund of Göteborg University spends much of her time sensitizing technocrats at Sweden's state utility, Vattenfall, to aesthetic concerns. Danish landscape architect Frode Birk Nielsen brought

20 years of experience with wind energy landscapes to the discussion. Most recently Nielsen has been applying visualization techniques to simulate the landscape impacts from some of the projects that will be necessary for Denmark to meet its ambitious renewable energy target.

The debate about the role of wind energy on the landscape in Great Britain has often been divisive and bitter. Wading into this controversy, Laurie Short has taken on the role of mediator through his Visual Arts Development Agency. He is personally familiar with the urban use of the rural landscape and has been instrumental in stimulating interdisciplinary discussions of Britain's countryside, including reactions to the introduction of wind turbines into the rural landscape near his home in Cumbria.

Two participants journeyed to Bellagio from Germany, the current world leader in wind energy development. Christoph Schwahn, a landscape architect, conducted one of the first studies of the influence of the then-new concept of wind farms on the flat polders of northern Germany. Many wind turbines are now concentrated on the reclaimed land he once studied. Today he can watch wind turbines sprouting from the countryside near the university town of Göttingen where his architectural practice is located. Martin Hoppe-Kilpper, an engineer, is at the center of German analysis of wind energy's technological success. As director of a wind energy program centered in Kassel, he literally monitors the performance of thousands of wind turbines across Germany in a federally funded program.

Dutch industrial designer Rob van Beek completed the team. Unlike the other participants, who are more at home writing articles and reports, van Beek's work actually appears in the design of wind turbines, including brightly colored examples on polders in north Holland. Van Beek has experimented with unusual painting schemes to accentuate the vanishing point along rows of wind turbines. He has also visualized unusual arrays of turbines. One novel circular alignment for a hypothetical wind farm he dubbed "windhenge," draws on wind energy's ecclesiastical or mystical overtones. Though he was unable to participate in this book, van Beek's views sharpened the discussion at Bellagio.

All participants brought to Bellagio not only their experience of working with wind energy but also a written presentation of their views. The papers collected here were honed at Villa Serbelloni. Although the Villa's surroundings were designed to induce harmony, the discussions were anything but harmonious. They sometimes were heated—quite heated. As a consequence, the voices presented here are not always in agreement, reflecting the ongoing conflict between convenience and cost, livelihood and landscape, nature and need.

NOTES AND REFERENCES

1. Additional criticisms about wind power include its hazard to birds, noise, and electromagnetic disturbances, but to date the evidence suggests that these are minor problems, compared to visual aesthetics, which will have small influence on wind power's future potential.
2. Comments by the staff of Danmarks Naturfredningsforening in an interview by Paul Gipe, Copenhagen, November 1997.
3. Bundesverband Landschaftsschutz, or BLS, is frequently mentioned in the pages of *Neue Energie* as opponenets of wind development. *Neue Energie (New Energy)* is the monthly news magazine of the Bundesverband Windenergie, the German wind turbine owners association.
4. Otfried Wolfrum Windenergie: Eine Alternative, die keine ist, as cited in Franz Alt, Jurgen Claus, and Herman Scheer, editors. *Windiger Protest: Konflikte um das Zukunftspotential der Windkraft* (Bochum German: Ponte Press, 1998).
5. 1 terawatt-hour = 1,000,000,000 kilowatt-hours (kWh) = 1000 million kWh or 1 billion kWh in American usage. One 100-watt light bulb operating for 10 hours will consume 1000 watt-hours or 1 kWh. One kWh produced by a wind turbine is the same as that produced by a conventional power plant.
6. 1 megawatt = 1000 kilowatts (kW). The kW and the MW are units of power. The size of power plants is given in kW or MW. This is the amount of power the plant can produce at peak production. Unlike many conventional power plants, which operate near their "rated" capacity, wind turbines operate at peak power only a portion of the time. The amount of energy delivered by a wind turbine for a given unit of power is often less than that from a conventional power plant. Thus, 1 megawatt of wind power is often less than 1 megawatt of a conventional power plant in its ability to generate electricity.
7. Robert W. Righter, *Wind Energy in America: A History* (Norman: University of Oklahoma Press, 1996).

PART

II

WIND POWER ON THE LAND

1

EXOSKELETAL
OUTER-SPACE CREATIONS

ROBERT W. RIGHTER

*Driving through Altamont Pass with the setting sun
over your shoulder, you see opening up before you the
vast Central Valley of California, and with luck the
serrated crest of the Sierra Nevada mantled in snow. In
December 1969 one small patch of ground near this pass
swarmed with 350,000 people attending an infamous
rock festival. Then came relative calm for 15 years.
Nowadays, all has been changed as a result of thousands
of wind turbines scattered over the site where the Rolling
Stones once played and far off in all directions. Although
the turbines unexpectedly protect the hills from suburbs
creeping in from both sides, equanimity has not become
part of the new scene. Writing from the perspective of a
historian, Robert Righter introduces the aesthetic context
of present-day wind developments that early citizen
reactions at Altamont helped produce.*

From Mount Diablo, it is said, one can see more of California than from
anywhere else in the state. Across the great Central Valley, the Sierra
Nevada creates a serrated horizon. To the west is San Francisco Bay; to the
south, Livermore Valley and Altamont Pass. In the early 1980s, the winds
that had been blowing invisibly through the pass thousands of years before
Sergeant José Francisco de Ortega first spied Mount Diablo in the 16th
century were suddenly manifest in the rotating blades of thousands of
modern turbines. Almost as quickly the turbines faced withering attacks
from those determined to maintain the hilly grass-covered charms of old.
Sylvia White, a professor of regional planning, expressed the views of
many when she accused wind energy companies of "industrializing" the
Altamont hills. The bucolic landscape, she suggested, had been made ugly
by thousands of spinning intruders. Professor White described them as

Copyright © 2002 by Academic Press.
All rights of reproduction in any form reserved.

FIGURE 1.1   Wind turbines lining ridge tops in The Altamont Pass, with fog rolling in from San Francisco Bay to the west. (Copyright Robert Dawson. Used with permission.)

"exoskeletal outer-space creations" with grotesquely anthropomorphic characteristics such as "long, sweeping blades attached to what ought to be their noses...[with] legs...frozen in concrete, stationary but seemingly kinetic." For White, "once-friendly pastoral scenes now bristle with iron forests"[1] (Figures 1.1 through 1.4).

Another challenge came from Mark Evanoff, head of the People for Open Space/Greenbelt Congress, an organization committed to maintaining a swath of open space encircling the great San Francisco/Oakland metropolis to the west. Within that green belt the Congress encouraged agriculture, wildlife habitat, and watershed preservation. To accomplish that goal Evanoff's group opposed the spread of suburban housing. More to the point, it opposed the proliferation of wind farms, seeing little

FIGURE 1.2   Wind turbines erected on hilltop behind preexisting ranch, The Altamont Pass, in 1986. (Courtesy Martin Pasqualetti.)

FIGURE 1.3   Wind installations in The Altamont Pass, looking northeast toward Stockton, California, in the San Joaquin Valley portion of the Central Valley in 1998. (Courtesy Martin Pasqualetti.)

FIGURE 1.4   Compatibility of ranching and wind generation, The Altamont Pass gives landowners a double source of income. These Danish turbines, installed in 1986, have since been replaced. (Courtesy Martin Pasqualetti.)

compatibility between spinning turbines and Congress objectives. Evanoff used language similar to that used by opponents of nuclear power, proclaiming that "we eventually will have to decommission the wind-mills."[2] For both Sylvia White and Mark Evanoff the wind turbines were industrial culprits, criminals that imprinted a rural environment with the gear of technology. "The greenbelt," Evanoff would say, "is not the place for light industry."[3]

The views of White and Evanoff have been echoed by many others in the United States and, I might add, in Europe. The worldwide use of wind energy, although expanding, has been slowed by the concerns of average citizens, often spearheaded by environmental groups. We do know that the NIMBY (not in my backyard) response is alive and well. Communities of people, often living near existing and proposed wind farms, have sharply voiced their opposition. Californians at Tejon Pass and Montanans at Livingston, for instance, have rejected proposed wind projects on the basis of the desecration of the landscape.[4] Sometimes these statements come from those one expects to favor its development. It is ironic that even such environmentally friendly methods of creating electricity nevertheless stimulate opposition within the community of environmental activism, but such a condition illustrates the complexity of the issue at hand and the wide range of public opinions that exist. Part of the problem seems to stem from a presumption of support from the community at large; in this sense, it would appear that wind energy planners have been intent on maximizing wind resources with insufficient consideration of the importance of public input. My goal here is to briefly examine both historic and contemporary attitudes toward wind turbines with the hope that those in positions of power will broaden their perspective, particularly with regard to the inclusion of public opinion.

THE NEED FOR WIND ENERGY

Wind energy is too abundant and thus too valuable to ignore. In the preindustrial past, humans did not always waste the wind. Worldwide, civilizations depended on water, wind, animals, and human muscle to accomplish necessary tasks. Even in the initial years of a new century we tend to overlook the potential of wind, treating it as neutral or an annoyance rather than a resource. The industrial world continues to rely on oil, natural gas, coal, and uranium. No one need be told that these sources are finite. Even if petroleum supplies were to prove unlimited, it makes little sense to continue its profligate and wasteful use. As with all natural resources, wisdom suggests conservation, particularly if we accept the evidence of global warming. One alternative is to increase the human use of kinetic sources of energy.

Throughout the past century the United States has developed its hydropower capacity. Workers constructed colossal dams throughout the nation, but particularly in the American West, where they not only generate electricity, but store water, a scarce resource. However, there are few sites left for large dams, and cost–benefit ratios at these locations are not favorable. Even if they were, environmentalists would fiercely defend the remaining free-flowing rivers. Today, even the Bureau of Reclamation, the dam-building arm of government, has acknowledged that its construction days are over. The agency must now focus on water conservation and water quality issues. Realistically, we can expect but few additional hydropower kilowatts. There are even plans now being formulated to remove some of the dams already in place.

In contrast to hydropower, wind energy is a rediscovered resource. Because it is diffuse, erratic, and uncontrollable, early 20th century engineers cast it aside. Few Americans, save a handful of sailors, thought of the wind as anything but an annoyance, at times a danger, and occasionally a destroyer. But in the past two decades it has become a deliverer, providing a fraction of American electrical needs. Ironically, a preindustrial energy source has found a place in the postindustrial world. Wind energy has made seven-league strides in the past 20 years. Engineers have made great improvements in efficiency and in reliability. The public has begun to take notice.

OPPOSITION TO WIND TURBINES

The question being raised in many countries — especially in the United States and in Europe — is whether there will be a land base upon which to place the new and improved turbines. Are Americans and Europeans willing to allow the intrusion of technology on cherished landscapes for the benefits of electricity produced in an environmentally friendly manner? Landscape architect Robert Thayer identified the issue: "Today we find ourselves in a deeply fragmented situation where we love nature but depend on technology."[5] Resolution will not be easy. As noted, the nation no longer believes that developing large-scale hydropower projects is worth the environmental cost. Planners and engineers, hard put to find a benign way to produce the electrical energy we want and need, may face a similar situation with wind energy. Already, many persons believe that the loss of pristine landscape is a sacrifice they are unwilling to make.

No one can provide wind developers with easy answers or formulas to overcome visual objections. They do not exist. Individual reaction to landscape and landscape change is complex. The geographer Yi-Fu Tuan suggests that each person will react to the physical environment, or nature if you will, differently. These differences may be attributed to body type, education, individual preferences, temperament, sex, and age.[6] Obviously, total public agreement on any project will be difficult, indeed impossible. If we are to concur with Tuan, not only culture but *individualism* will complicate the task of a planner's effort at consensus. If our response to wind turbines is prompted by individual preferences rather than cultural influences, we will each react differently to wind turbines placed on the landscape. Of course, realistically we cannot ignore cultural influences or individual preferences. Reaction to landscape intrusion is a *blend* of both.

ORIGINS OF OPPOSITION

Although one must acknowledge that individualism will dilute the influence of culture, understanding our American heritage can offer some perspectives on the dilemma. The paradox of our love of nature and our dependence, perhaps worship, of technology has resonated in the American character since the time of independence. The practical value of technology has almost always won out.[7] Intellectual historian Leo Marx underscored this long-standing conflict between our love of nature and technology in his book *The Machine in the Garden*.[8] While some Americans, perhaps best represented by Henry David Thoreau, questioned the shrill whistle of the locomotive, most citizens welcomed this dominant 19th-century technology. When Thoreau pronounced that he wished "to speak a word for Nature, for absolute freedom and wildness," few Americans cared or understood his message.[9] They welcomed the coming of the railroad and the consequent creation of a pastoral landscape in place of wilderness. Yet a few such as Thoreau and the social reformer Henry George weighed the benefits against the costs, and then stated their objections, albeit based on economic rather than environmental principles.

The "iron horse" was not welcomed by all. New technology has always been suspect, and often opposed. This was even true of windmills. Although we often think of the English post mill and the Dutch windmills as intermediate technology in harmony with the surrounding countryside, these windmills were not without their critics.

Protests took varied forms, and they have not always been based on visual objections. For instance, in the 1180s, Abbot Samson, the dictatorial head of Bury Saint Edmunds in Suffolk, England, went into an almost crazed rage when Herbert, an adjacent land owner, erected a post mill. Samson saw to it that the mill came down, but not for aesthetic reasons. He simply refused to condone grain-grinding competition for his nearby watermill.[10] In essence, those who controlled England's water power wanted no competition from the wind.

Windmills were also suspect because of a general reputation of millers at the time. They were not trusted by the provincial population, for they often "adulterated meal and finely ground flour with powdered bark and roots, with ground limestone, and with sand, returning to hapless peasants not the wholesome fruit of agricultural labor but an artificial mixture fit for no man."[11] Furthermore, the miller often kept more than the one-fourteenth grain fee to which he was entitled, or at least he was *suspected* of such chicanery. Millers were essential, but not necessarily popular, and by association, neither were their windmills.

Laborers have often opposed new technology because it could lead down a path to unemployment. In 1768, for example, workers in the sawmill town of Limehouse — near London — complained over the construction of a windmill. Their protests went unanswered and, fearing the loss of their jobs, they destroyed the offending windmill.[12] Again, the basis of their protest was economic, not aesthetic, though one suspects that in the eyes of these working people, a whirling windmill was *not* a thing of beauty (Figures 1.5 and 1.6).

Thus, contrary to accepted opinion, the old mills were not universally loved. And, of course, since they represented new technology, those persons who lived and worked with the mills had no nostalgic feel for them. Nostalgia is a product of time and age, and perhaps a distancing from the technology in question. We have a sense today that the windmills of old fit the landscape because they employed technology at what we perceive to be an acceptable level of disruption to landscape and nature. But this is hindsight. If they fit so perfectly, no one *at the time* would have had negative comments, either economic or aesthetic, toward English or Dutch windmills, icons of beauty and quaint technology today.

FIGURE 1.5   Windmills typical of the hundreds of thousands that used to dot the Great Plains, now preserved at the American Wind Power Center, an outdoor museum in Lubbock, Texas. American water-pumping windmills, such as this, were scattered across the countryside at individual farmsteads, unlike the concentration of modern turbines in sometimes large clusters. The interest in preserving the traditional American farm windmill contrasts strongly with the inclination by some to oppose installation of more modern designs. (Courtesy Martin Pasqualetti.)

FIGURE 1.6   A 1900 Model Duplex Vaneless water pumper at the American Wind Power Center. (Courtesy Martin Pasqualetti.)

It would be foolish to argue that the windmills of the past faced the degree of opposition that today's huge turbines encounter, but it is true that throughout history the introduction of new technology of any sort has met opposition. Usually opponents have structured their arguments on economic issues, but certainly social and political concerns were not absent. Neither were environmental considerations. Fear (nuclear reactors), health (lead smelters), smell (feed lots), and noise (airports) are common environmental catalysts for community resistance. And, of course, so is visual pollution. Movement, such as that of a whirling turbine, may not evoke the same public fear as a nuclear accident, but many Americans find it a distraction and an annoyance. Such sentiment, when a cherished landscape is being affected, is intensified.[13]

More than a century ago the issue between the aesthetic and the utilitarian was not altogether different than today. Even the American windmill, essential in the development of the western grasslands, had its detractors. In 1886, one Chicagoan complained that lands to the west were "dotted all over with unsightly patent windmills used for pumping water, generally further disfigured with the name of the particular make of the windmill." This observer did not want them torn down, but he did suggest change. He noted that "there is no need . . . that these useful appliances should be as ugly as in most cases they are." With proper design and a little extra cost, they might be made "exceedingly picturesque, and add to the interest of the landscape...."[14] A reply by the *Chicago Tribune* opened with: "Never mind the esthetics." Stressing the utilitarian, the newspaper noted that the cattle who drink the well water "are quite satisfied if the mill does not look quite as nice as some people would like."[15]

DEFINITIONS OF LANDSCAPE

Human beings are sensitive to landscapes. Perhaps all people acknowledge the beauty of natural landscape. Many see it as central to their lives, and some invest it with divinity. With such an emotional attachment, little wonder that people object to the altering of that environment. But what is landscape? The definition of landscape is central to all discussions about wind power aesthetics. My own orientation is historical and confined to North America. Landscape architect John Stilgoe claims that "a landscape happens not by chance but by contrivance, by premeditation, by design; a forest or swamp prairie no more constitutes a landscape than does a chain of mountains. Such landforms are only wilderness, the chaos from which landscapes are created by men intent on ordering and shaping space for their own ends."[16] Perhaps people feel strongly about landscape because we have an emotional investment: we created it. If this is true, then perhaps we can more readily accept wind turbines on land in which humans have had little impact. The desert lands of the West are most obvious. Here humans have in many places invested little labor, and it can be held, neither has nature.[17] It should come as no surprise that wind farms have found homes in the desert West.

The late J. B. Jackson strengthens Stilgoe's definition of landscape. He asserts that "landscape is not scenery...it is really no more than a collection, a system of man-made spaces of the surface of the earth." Jackson believes that the natural environment "is *always* artificial": that is, created by people.[18] I must respectfully disagree. We do draw artificial boundaries for wilderness areas, but within those boundaries we intend to turn over ecological responsibility to nature, or natural processes. Some of the most unforgettable landscapes etched in this writer's memory have been in the high lake regions of the Wind River Range and the long, mountain-encircled meadows of the Sierra Nevada Range. Such lands have a simple purity because human manipulation is absent. Perhaps we need a new word, such as "wildscape."

Historically, of course, Americans have been ambivalent regarding wilderness lands: some equated such lands with danger (Indians, grizzly bears, starvation, disorientation), where others saw them as an inviting alternative to corrupt civilization.[19] Today, pleasure has replaced fear, and our exploitive urges have been tempered with respect and, indeed, awe. No one who has spent time in the wilderness would condone the violation of that landscape by wind turbines. No matter what the wind resource, it is off limits to human exploitation.

But let us return to the Stilgoe and Jackson presumption that landscape is a human construct, essentially the manipulation of nature to suit our aesthetic or economic needs. Even though it is the product of deliberate human intervention, we do not like to admit it. The most desirable landscapes are those which give little evidence of human management. For many of us, the helter-skelter look of an English country garden is preferable to the formal gardens of Versailles. Although both were created through the imagination of landscape gardeners, the English garden fools us into believing that it is more "natural," that nature had a significant hand in its creation.

The desirability of natural landscapes has made the English countryside world renowned. Winding roads, curving fences, wooded hills, hedgerows, and green pastures combine to render our ideal of an aesthetic landscape. The combination of natural elements gives the illusion that nature does the planning here. Obviously, there is human order, but it is subtle, hidden from view. What does come to mind is harmony: an appealing symbiosis between people and nature. It is as if people and the land have coexisted here in the past, in the present, and will continue in the future. Human alteration of this landscape harmony will have to be slow, almost imperceptible, simply because we all suffer from what Robert Thayer defines as "landscape guilt": that is, the premise that "Americans, in increasing numbers and intensities, feel guilty about what technological development has done to the landscape, to 'nature,' and to the earth."[20]

If Thayer is correct, and I believe he is, this is not good news for the wind turbine business. While many Americans acknowledge that the wind is free and benign, they also realize that the harvest of this resource is highly technological. But it is not the technology so much as the *visibility* that people find objectionable. Engineers cannot hide wind turbines. They cannot camouflage them. Neither can they build transparent or invisible wind turbines. We cannot shield the public from their presence.[21] Engineers can erect them in isolated places, and this has surely happened, such as the array of Micon turbines on Southwest Mesa, near McCamey, south of Midlands, in West Texas (Figure 1.7). Yet, wind energy should not be relegated by default to the most desolate places in the nation, far from transmission lines and consumers. Such a fate would surely limit wind energy, particularly in Europe and the more densely populated regions of the globe.

FIGURE 1.7   Part of an array of 107 wind turbines installed at the Southwest
Mesa Wind Project, near McCamey, in West Texas, dedicated June 1999 by
Governor George W. Bush. Together the 700-kW NEG-Micon generators have a
capacity of 75 megawatts, making it at the time the largest wind power
installation in Texas. It is located amid one of the most productive, but now
largely spent, west Texas oil fields. (Courtesy Martin Pasqualetti.)

HOW TO INCREASE PUBLIC ACCEPTANCE

But what can landscape architects, engineers, developers, and even
historians do to increase public acceptance? All of us must educate the
public about the environmental benefits of wind energy. Beyond that, we
at Bellagio believe that the landscape architect must employ his or her
skills in seeking compatibility between nature and this technology, while
the engineer must create designs which are reliable, yet more pleasing to
the eye.

As mentioned earlier, the historian can offer the perspective that
technological change has never been universally welcome: not electricity,
not automobiles, not the airplane, nor nuclear energy. Wind energy can
expect no less skepticism. What is evident, culturally speaking, is that
wind technology is on solid ground. For example, in the book of Genesis
God's people were given dominion over the earth, and in that gift there
were no strictures against the use of tools (technology) to establish that
supremacy.[22] Politically, the same may be true. One is reminded of the

debate played out by two of the "Founding Fathers": Thomas Jefferson and Alexander Hamilton. Jefferson promoted the nontechnical, yeoman-farmer lifestyle. He favored a pastoral way of life, featuring a life close to nature. Alexander Hamilton, his nemesis, supported industrialization, the city, commerce — and technology. By the close of the Civil War, long after both men had died, Hamilton's view had clearly won out. The industrial North had defeated the rural South, ending not only slavery but agrarianism as the major occupation of the nation. From the Civil War on, America never reversed its commitment to technology, industrialization, and urbanism.

With regard to landscape patterns (taking the view that landscape is a human construct), the marriage of technology and nature came early in American history. Ironically, Jefferson had much to do with it in his fashioning of the Land Act of 1785. In this legislation topophilia (love of land) and technophilia (love of technology) were both much in evidence. Contrary to a land ownership system featuring the use of natural landmarks, such as a stream, the dividing hill of a watershed, an outcropping of rock, or a unique tree, the Land Act mandated the scientific surveying of the land in precise tracts. The act determined how Americans would define, measure, appraise, and sell land. It did not impose any zoning restrictions, except in the reservation of land for schools. But above all, it imposed a technical method of defining land parcels through surveying. Consequently, township, range, section, and quarter-section are now part of the American land lexicon and the landscape itself.

When we examine American constructed landscapes we are struck by the preponderance of squares and straight lines, and the neglect of natural features, monotonous in its "disregard for the topography."[23] Nature does not have straight lines in its aesthetic storehouse of options. Natural landscapes feature both vertical and horizontal irregularity. But back in 1785, a barely independent nation changed all that. Government surveyors would transform the formless American wilderness into what one historian described as "a remarkable national geometry of gigantic squares and rectangles varying from 640-acre sections to 23,040-acre townships."[24] Over the years the lines and squares would shrink into 160-acre homesteads, or even smaller 80- or 40-acre parcels. But no matter what the size, the geometric pattern remained. Any person who has flown across the American Midwestern states on a clear day cannot fail to be struck by this orderly layout of thousands of squares and rectangles which altogether ignore natural barriers. The United States has never legislated a national land use plan, but the Land Act was close. It arbitrarily dictated that the national landscape would be one of order, of squares and straight lines.[25] Certainly it represents Enlightenment ideas regarding order and rationality, but above all, it is utilitarian. It is a gridiron landscape which ignores natural features, all in the name of efficiency and utility. In our cities as well, the gridiron pattern prevails, often in contradiction to nature and creativity.[26]

Thus from the beginning, the United States seem to disavow the importance of natural landscapes. Can our past tendency to view land in orderly geometric patterns tell us anything about the placement of wind turbines on the landscape? I think so. I believe that frequently we see a similar pattern. Visitors to the passes of Altamont, Tehachapi, or San Gorgonio landscapes will often find the straight lines of land surveys replicated by the rows of turbines. These rows are placed, presumably, to impose order: the same order on the wind as the geometric survey imposed on the land. However, as with land, the straight lines of turbines do not enhance the natural landscape, but merely emphasize the heavy hand of utilitarianism.

Such orderly development is neither attractive nor inviting. Straight lines connote artificiality, the antithesis of nature. For example, New Zealand foresters make no attempt to hide the fact that they are farming trees. The pines are planted in long lines, not much different from rows of corn or Christmas trees. These forest landscapes neither inspire nor do they draw hikers or picnickers. Such forests have become constructs of man, commodities if you will. Lines of wind turbines are not altogether different. Their distribution repels rather than attracts. Because they are steel and they are massive, the long lines appear to be a stationary, yet moving, army. Had Don Quixote been battling lines of turbines, history might have judged him quite sane! And, of course, there are many modern-day Don Quixotes who are more than ready to take up the lance. They are struck with severe cases of technophobia. Critics such as Sylvia White and Mark Evanoff have their supporters who find these rows of metallic, robot-like machines ugly, the antithesis of nature and naturalness.

VIEWS OF TURBINE PLACEMENT AND DESIGN

Wind energy consultants must look to placement in order to facilitate landscape compatibility. For instance, long lines of turbines offend many viewers. However, is a clustered placement more natural, and hence, more acceptable? Could an arched or curved row be more inviting? Today farmers in some areas have broken up the tedium of rectangular survey lines by center pivot irrigation systems. Admittedly such change is for economic reasons, but still, the square becomes a circle. Perhaps this may provide an option for wind turbines? What consultants must remember, as Yi-Fu Tuan tells us, is that we are all individuals with different perceptions of landscape.[27] I, for one, prefer the clustered placement of Kenetech wind turbines on Delaware Ridge in West Texas rather than the army-like rows of turbines at San Gorgonio Pass. However, perhaps my choice is based on a preference for hilly rather than flat topography.

Not only must placement be a factor in wind farm development, so must turbine design. At present there would appear to be only one reliable turbine, the Danish-type three-bladed machine. However, such dominance of one style limits choices. The Danish design may be the most efficient, but is it always the most harmonious design for the landscape? Perhaps not. In certain landscapes the vertical-axis Darrieus rotor may be more pleasing to the eye. In some situations a slower turning machine may attract the eye rather than repel it. Some of the more bizarre designs of the past may offer inspiration. Recently a young designer contacted me, seeking more information on Dew Oliver's "blunderbuss" design of the 1920s. It had a low profile and a minimum of movement. It may be a very acceptable alternative in some situations. Will it work? It did for Dew Oliver. I'm pleased that an engineer is exploring, or at least thinking, about the possibility.[28] Such creativity should be encouraged. Research and development have proven that other designs are efficient and may be more in harmony with the landscape, and yet these designs go unperfected. More attention must be paid to the aesthetics of wind turbines. We need world competition in wind turbine design, and the winners should not be judged *only* on efficiency, but also on compatibility in different landscapes. One size (style) does not fit all (environments). Rejection of a wind energy site should be a trumpet call to creativity and innovation, not to confrontation and discord.[29]

WIND TURBINES AS ART

I have made an assumption with which many will disagree. That assumption is that wind turbines should be in harmony with nature, that they should blend, and be as unobtrusive as possible. Some will say no; individualism is the master when subjective judgments rule the query. In my own limited experience, many observers of the large California wind farms either love or hate them. Few are without an opinion. Many find them absolutely fascinating, surrealistic in their transformation of the landscape into an artistic, futuristic, human-controlled spectacle.

One is perhaps reminded of the sculptor Christo and his artistic expressions strewn, as some would say, across the American West. Of particularly interest was the "Running Fence" in the 1970s. Christo and his army of volunteers erected a broad sail-like sheet across some 20 miles of rolling hills in Marin County, California, ending in the crashing waves of the Pacific Ocean.[30] Opinions were widely divided on this project: many deplored the expense and senselessness of the undertaking, while others praised the whimsical, creative, nonfunctional aspects of the fence. I claim little knowledge of Christo, but his work does alter nature with art. He makes no effort to hide this fact. His work transcends the landscape as the hills, canyons, and grasslands become the vast stage for his creation.

Closer in design to wind turbines have been Christo's umbrella projects. In both Japan and along California's Interstate 5 freeway near Tejon Pass and the "Grapevine," Christo erected hundreds of colorful yellow umbrellas (blue in Japan). Placed randomly, in some respects they do blend with the landscape. However, the vibrant colors beg to be noticed, and clearly they are meant to relegate nature to a mere backdrop. "The project is really about art," stated Christo. As usual some agreed, others did not. Perhaps the most germane comment came from an art critic who believed that the project "shaped nature and in such a brilliant way." Another waxed eloquent, proclaiming that "the umbrellas surprise and refresh our eyes, reawakening them to the beauties of that sere and inhospitable terrain." Seen from a distance they were evocative of California's golden poppies; close by, "small temples."[31] For some observers, the 1340 yellow umbrellas enhanced and improved a rather barren landscape.

Of course, Christo's creative works have been temporary, constructed one month and deconstructed the next. Yet the acceptance of such a dramatic alteration of landscape, even on a temporary basis, indicates that for a portion of the population, the imposition of art with technology is an acceptable modification. Even if such modification is unacceptable, we may not have a choice. Most of us live in an urban landscape, where change is pervasive. Transformations may sadden us, but we accept them as inevitable—and sometimes even positive. If we can live among the constant alteration of our cityscapes, the same may be true of our more natural landscapes.

BEAUTY AND ARCHITECTURAL BEASTS

Paul Gipe gives us a particularly pertinent example of how the public reacts to new objects in their environment: in this case, Paris. Parisians were initially shocked and repulsed by the erection of the Eiffel Tower in the center of what is arguably the most beautiful cityscape in the world.

Today the tower is a remarkable success, and when I visited recently, workers were busy maintaining and servicing it. I expect they hope it will last forever. Certainly those who ridiculed the tower are forgotten in history, and French school children today learn that the historic battle to save the tower symbolized "public-spirited perseverance surmounting narrow-mindedness, fear, and intolerance."[32]

If we are as flexible with landscape as the Parisians were with their cityscape, perhaps there will be minimal opposition to future placement of wind turbines. A new generation adjusts, and what may be offensive to the old is pleasing to the new. I have written earlier in this paper and elsewhere that the 500-kW turbines in mass "evoke feelings of technophobia. They are steel and they are massive [and]...they seem to rival nature rather than cooperate with it."[33] Not everyone agrees. Based upon his surveys, primarily conducted in the mid-1980s, Robert Thayer considered such a judgment unnecessarily pessimistic. At Altamont Pass, which Thayer described as a "highly conspicuous, man-made landscape development causing widely mixed reactions among viewers," he and associate Carla Freemen found that reactions to the wind turbines were varied. One subject wrote: "I truly appreciate the fact that windmills can offer a safe addition to the already available energy sources. I was extremely disappointed at the way the windmills distract and disturb the local environment. But with choices that are available today that disappointment has eased a little."[34]

KEEP THEM TURNING

As evident in the foregoing response, people are ambivalent toward the established wind networks. However, the public's response is heavily swayed by the *condition* of the turbines. Wind machines must be in good repair and functioning. It is essential, and perhaps the most important aspect of public relations. If the blades turn they confirm the public's expectations of environmental benefits. Unfortunately, in the 1980s broken blades and stilled machines all too often gave the California traveling public the impression that they had just passed through a wind energy cemetery. Many felt that they had sacrificed a rural landscape for a technological graveyard and an artistic calamity. This was not an attractive landscape. Tuan notes that landscapes such as "the denuded hill country of South Carolina with bed springs and tin cans in its gullies" can be demoralizing and a symbol for a defeated past.[35] So can defunct windmills. However, to set this impression on its head, can we say that a field of wind turbines, properly maintained, is a landscape of a hopeful environmental future? It is a landscape where the negative industrializing effect is outweighed by the positive environmental benefit. But, again, the turbines must spin. When they do, for many the beauty is in the utility of benignly created electricity.

Fortunately, the reliability of new, third-generation wind turbines is gradually winning back supporters. However, history counsels us that the sins of the past are not easily forgotten. In the early 1980s profit-minded operators put up shoddy turbines, thus ruining the reputation of a nascent industry. New projects must face a residue of public mistrust and dissatisfaction. Furthermore, large corporations such as Boeing received lavish federal funding to develop large turbines, but produced only failure and left the field when the money dried up.[36] We can hope that this episode will not be repeated and the trash of the past will be removed completely from view.

THE PUBLIC IS FORGIVING

Every technology has had its experimental period. There was a time in aeronautical engineering when planes crashed more than they flew. Robert Thayer and others believe that the public is forgiving, and that ambivalence can be transformed into support, if energy developers make their case in a sensible way. He believes that well-designed and well-sited wind energy projects can achieve a serviceable beauty common to other working landscapes. Such optimism is refreshing. I have been more in the skeptical camp, believing that this industrialization of the landscape will not be acceptable to the American public. However, we are very capable of change. A few years ago I wrote Thayer to express my belief that technophobia would endanger future projects. He responded that intrusion on the visual landscape was being countered "by an accrual of positive environmental symbolism."[37] There is some evidence that he is right. Rotating wind turbines have emerged as popular icons in Hollywood films and television advertising — often symbols of progress, modernity, reliability, and environmentalism. They are often juxtaposed with quality automobiles, reliable airline companies, and futuristic computers.[38]

THE POWER OF DEDICATED OPPONENTS

Even if a majority of the public clearly favors a wind energy project, that is no guarantee that it will be approved. Small numbers of dedicated opponents can and will attack projects, crushing developers with their passion. In both Tejon Pass, California, and Livingston, Montana, wind projects were defeated by relatively small clusters of adversaries, people usually directly affected by the proposed projects. Although the great majority of persons might approve of an *idea*, a modest number of opponents can defeat a specific project.

Some activists expect to profit by land development which a specific wind project would threaten, but the most ardent have emotional attachments to the land in question. By *attachments* I mean that they have a "sense of place" regarding the site which is to be altered. Jackson tells us that this "sense of place" entails a "certain indefinable sense of well-being which we want to return to, time and again."[39] Obviously this attraction is connected to memory, a powerful emotion. Tuan notes that "the appreciation of landscape is more personal and longer lasting when it is mixed with the memory of human incidents."[40] People who have memories attached to landscapes do not want change. Is it safe to say that we all have special places — a park, a canyon, a vista, an area, a rural setting — that we return to regularly, if only in our memory? The possibility that special places may be visually altered by hundreds of wind turbines will trigger determined opposition. Perhaps documenting the obvious, Thayer and Freeman found that opposition at Altamont Pass was strongest with those living close to the area.[41] Familiarity with a place generates attachment, and indeed love, of that landscape. With love comes a sense of stewardship and a determination to protect the land as it is. Wind turbines can be anathema to that purpose.

Of course some landscapes inculcate that "sense of place" more than others. Yet, even the desert, the receptacle for much of human refuse of one sort or another, will have its defenders — those persons emotionally attached to preservation. Wind energy promoters can expect to find some opposition to almost any site on land or offshore.

A WORD FOR THE INDUSTRY

A decade ago a consulting group known as Future Technology Surveys brought together 17 wind energy experts. Among the 17 were 3 CEOs, 3 vice presidents, 3 engineering managers, 7 researchers, and 1 marketing manager.[42] The survey personnel asked the experts a number of questions concerning the future of the wind energy business. "What are the most significant barriers to entry for new firms...?" "What technological pitfalls do you foresee for the wind power business?" "What specific developments do you foresee occurring...?" Responding, the 17 experts paid little attention to environmental problems, focusing on those of economics and engineering. They did acknowledge that "environmentalists [who] do not like windmills 'cluttering' the landscape" could be a barrier. Furthermore, they suggested that "site pollution due to moving blades, high tower[s] and blade noise" could be a technological pitfall.[43]

Perhaps the most revealing data came with the question: "What do you see as the greatest research needs related to wind power?" The group listed 24 areas, among them manufacturing cost reductions, energy storage, improved towers, blade technology, control systems, and reliable gear trains.[44] *These experts did not list one environmental problem as worthy of inquiry.* Perhaps this group considered that only scientists can do research. Perhaps they did not consider that the expertise and investigative skills of landscape architects, geographers, philosophers, or even historians could be useful to them. The survey does confirm what Robert Thayer and Heather Hansen concluded: "Wind-energy developers have largely ignored the public sector and grossly underestimated the continued strength of public sentiment for the rural, pastoral settings that turbines eventually occupy."[45] It intrigues me that it never occurred to these executives who spend their lives and creative efforts promoting wind energy that they need help in understanding the public with which they must strike a compromise. Furthermore, they need assistance (read research) in figuring both how to design friendly turbines and then how to find acceptable sites in which to place them. Again, the engineers may build reliable, efficient wind turbines, but that is only a fraction of the solution. What if they are ugly and there is no place to erect them? I once wrote that "the future of the wind-energy field is a matter not only of engineering, but of the social sciences and the humanities. Many fields of knowledge must make contributions if barriers are to be overcome. The tendency of the engineering community is to knock them down, but it is time to consult those who would quietly dismantle the barriers brick by brick...."[46]

Now, more than a decade has passed since that 1991 meeting. Presumably, most engineers and planners now understand that a wind turbine is more than blades, gears, a generator, a nacelle, and a tower. It represents a visible addition to the landscape. It should transform that landscape into one of environmental hope, signifying to the public that there is a long-range future for humankind. It should, as well, recast the landscape into one of utility, but still one of—dare I say it—beauty and harmony. It should allow the individual to order his reality from different angles.[47] If engineers, designers, and landscape architects all do their jobs, is there a chance that a wind turbine landscape could inspire us as it did the 19th-century English writer William Cobbett? Early on a sparkling day, Cobbett looked down upon a valley ringed with 17 windmills: "They are all painted or washed white; the sails are black; it was a fine morning, the wind was brisk, and their twirling altogether added to the beauty of the scene, which . . . appeared to me the most beautiful sight of the kind that I had ever beheld."[48]

NOTES AND REFERENCES

1. Sylvia White, "Towers Multiply, and Environment Is Gone with the Wind," *Los Angeles Times*, November 26, 1984, Sec. II, 5.
2. As quoted in Seth Zuckerman, "Winds of Change," *Image* (September 20, 1987): 30.
3. Ibid., 29.
4. See Paul Gipe, *Wind Energy Comes of Age* (New York: John Wiley & Sons, 1995) 446–450, and Robert W. Righter, *Wind Energy in America: A History* (Norman, Oklahoma: The University of Oklahoma Press, 1996), 236–240, 251–259, for details on these two controversies. Visual pollution has not been the sole objection. Avian mortality and noise have also elicited opposition. Also see Peter Asmus, *Reaping the Wind* (Washington, D.C.: Island Press, 2001), 137–142, for more on Altamont and Tejon Pass opposition.
5. Robert L. Thayer, *Gray World, Green Heart: Technology, Nature, and the Sustainable Landscape* (New York: John Wiley & Sons, 1994): 94.
6. Yi-Fu Tuan, *Topophilia: A Study of Environmental Perceptions, Attitudes, and Values* (New York: Columbia University Press, 1974, 1990 ed.): 45–58.
7. There are exceptions. I am reminded of the defeat of subsidies for development of an American Supersonic Transport in the 1970s. Also, the Wilderness Act of 1964 represents a significant victory for nature. However, perhaps the exceptions do prove the rule.
8. Leo Marx, *The Machine in the Garden: Technology and the Pastoral Ideal in America* (New York: Oxford University Press, 1964).
9. Henry David Thoreau, *Walden and Other Writings*. Edited by Brooks Atkinson (New York: The Modern Library, 1950 ed.): 597.
10. Edward J. Kealey, *Harvesting the Air: Windmill Pioneers in Twelfth-Century England* (Berkeley: Univ. of California Press, 1987):132–153.
11. John R. Stilgoe, *Common Landscape of America, 1580 to 1845* (New Haven: Yale University Press, 1982): 300.
12. T. K. Derry, Trevor I. Williams, *A Short History of Technology* (New York: Oxford University Press, 1961): 257.
13. For example, I am rarely offended by visual intrusions in my Dallas environment, a cityscape. However, at my retreat in Jackson Hole, Wyoming, a similar intrusion would result in an emotional response.
14. "The Picturesque in Windmills," *The Farm Implement News* (Chicago), VII, No. 2 (February, 1886): 16.
15. Ibid.

16. Stilgoe, *Common Landscape*, 3.
17. I am aware that ecologists would vehemently dispute the idea that nature has not invested much effort in the desert.
18. John B. Jackson, *Discovering the Vernacular Architecture* (New Haven: Yale University Press, 1984): 156.
19. See Chapters 2 through 4 (pp. 23–83) in Roderick Nash, *Wilderness and the American Mind* (New Haven: Yale University Press, 1967, rev. ed. 1973) for an enlightening discourse on the duality of fear and love of wilderness in 18th and 19th century America. Stilgoe, pp. 9–12, underscores the fear of wilderness but underplays the romantic.
20. Thayer, *Grey World, Green Heart*, 55.
21. Our team discussions at Bellagio, and particularly the views of European landscape architects, have altered my thinking. By use of color and design it is becoming increasingly possible to reduce, but certainly not eliminate, turbine visibility.
22. The question of the Judeo-Christian heritage as a cause for the environmental crisis is one that has attracted a body of literature. Historians Lynn White and Roderick Nash and the geographer Clarence Glacken have investigated the subject. Landscape architect Ian McHarg and biologist Rene Dubos have added to the debate. This is certainly not the place to debate whether the Judeo-Christian heritage is one of dominion or stewardship; however, it is safe to say that within that heritage there were no strictures *against* the use of technology to control nature.
23. John Brinckerhoff Jackson, *A Sense of Place, a Sense of Time* (New Haven, Connecticut: Yale University Press, 1994): 4, 151.
24. John Opie, *The Law of the Land: Two Hundred Years of American Farmland Policy* (Lincoln: University of Nebraska Press, 1984): 1.
25. For an explanation of the Land Act of 1785 see ibid., 1–17, 44–56. In human terms, the tension between a natural land system and the geometric land survey was played out between the squatter (natural) and the land speculator (survey). The result was often violent and is a major theme in 19th century American land history.
26. Ian McHarg, *Design with Nature* (New York: John Wiley & Sons, 1995 ed.) presents an alternative, more natural approach to city planning.
27. Tuan, *Topophilia*, 48.
28. For more information on Oliver's design, see Righter, *Wind Energy*: 87–90.
29. On the question of design, the Bellagio team disagreed. Paul Gipe and Frode Birk Neilsen led a faction which argued that we must accept the Danish three-blade machine. Others, including myself, believe that continued research may reveal an efficient, more aesthetic design.
30. See *Christo: The Pont Neuf Wrapped: Paris, 1975–1985* (Bonn, Germany: Otto, 1994): 17–19.
31. See the *Los Angeles Times*, October 4, 1991, B1; October 18, F4; October 21, F3. Ironically, Christo's umbrella project ended rather disastrously when Lori Keevil-Mathews, a 34-year-old woman from Camarillo, California, was killed when a strong gust tore out one of the 488 lb. umbrellas and crushed her against a rock. A depressed Christo shut down the project shortly thereafter. *Los Angeles Times*, October 28, A1.
32. Gipe, *Wind Energy Comes of Age*, 252–254.
33. Righter, *Wind Energy*, 31.
34. Robert L. Thayer, Carla M. Freeman, "Altamont: Public Perceptions of a Wind Energy Landscape," *Landscape and Urban Planning*, 14 (1987): 393.

35. Ti-Fu Tuan, "Thought and Landscape," in D. W. Meinig, ed., *The Interpretation of Ordinary Landscape: Geographical Essays* (New York: Oxford University Press, 1979): 100.

36. I have often said: "Thank God that Boeing builds better airplanes than they do wind turbines!"

37. Robert Thayer to author, February 25, 1992. Also see Thayer, *Gray World, Green Heart*, 273–276.

38. TV advertisements for the Lexus automobile and U.S. Air, both featuring wind turbines, may be evidence of changing perceptions. Also, Compaq Computer Company has used wind turbines in advertisements featuring modernity and a futuristic outlook.

39. Jackson, *A Sense of Place, A Sense of Time*, 157–158.

40. Tuan, *Topophilia*, 95.

41. Thayer and Freeman, "Altamont: Public Perceptions . . . ," 379.

42. Richard K. Miller, Marcia E. Rupnow, *Survey on Wind Power*, Survey Report #124 (Future Technology Surveys, Inc., 1991): 6.

43. Ibid., 23, 26.

44. Ibid., 19–20.

45. Robert Thayer, Heather Hansen, "Wind on the Land," 72.

46. Righter, *Wind Energy*, 278.

47. I am paraphrasing Arnold Berleant, *The Aesthetics of Environment* (Philadelphia: Temple University Press, 1992): 63.

48. Quoted in Stanley Freese, *Windmills and Millwrighting* (Devon, England: David & Charles, Newton Abbot, 1971): xiv.

2

WIND POWER AND ENGLISH
LANDSCAPE IDENTITY

LAURENCE SHORT

*More than their poets, their art, or their architecture,
the English love their landscape, and woe betide any
who would threaten it. This protectionist attitude has
brought wind development in England nearly to a
standstill. Drawing from personal experience and
public discussions of wind power, Laurence Short here
argues that reactions to the changes that wind power
bring to the land range between the romantic urban
dwellers' remembrance of landscapes past and the views
of those living within the working countryside of the
present. He suggests that a widened public debate over
the future of wind power that is more holistic and
inclusive of the artist's perspective must precede wind
turbines becoming icons for a sustainable future.*

If the wind industry is to gain public acceptance, especially in England,
it must address community interests among a wide range of environmental
aesthetics, but above all the value of landscape to our culture. It must face
the growing conflict between nature and technology (Figure 2.1). And it
must do so, I contend, by making a case within the framework of a holistic
environmental approach. I suggest that the wind energy debate must be
expanded beyond monetary profits alone to include the value of the
public's perception of landscape. This voice and this message are simply
too important, too powerful, to be ignored.

The recurring resistance to wind energy is primarily one of public
relations, yet an important factor is being ignored: Both pro and con
positions have mutual ownership of the landscape. Wind energy leaders
fail to grasp the important links among landscape, memory, and beauty in
achieving a better quality of life. And particularly, they fail to realize that
these links are spiritually important not only to a rural populace, but to
urban people as well.

Wind Power in View:
Energy Landscapes in a Crowded World

Copyright © 2002 by Academic Press.
All rights of reproduction in any form reserved.

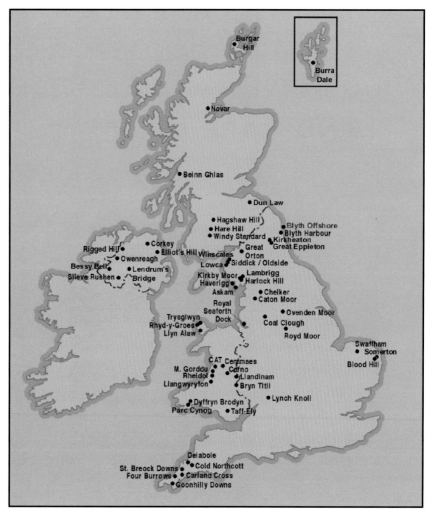

FIGURE 2.1   Distribution of wind installations in the United Kingdom. (Copyright British Wind Energy Association, http://www.bwea.com. Used with permission.)

If the wind industry continues to ignore the public's fear of environmental change, it will reap only harsh rewards. At present it oversimplifies landscape aesthetics when it asserts that all objections to projects are simple-minded reactions against *any* development whatsoever. Does it really believe that naming and shaming its adversary will be sufficient? Unfortunately, it often makes the fatal mistake of not examining the value and the interests of this seeming cloud of mosquitoes! Rather than recognize the nature of the problem, the industry not for the first time seeks to gain acceptance through improved technology! It opts for a more acceptable design or a larger turbine, a strategy that only reinforces the public's view of an aggressor rather than a collaborator.

To become a collaborator rather than an invader, wind energy developers must take a wide-angled look at the barriers to changing our perception of landscape. Those of us who understand the environmental value of wind energy must place the debate firmly into one which considers community, environment, and, indeed, survival. Survival is our common interest. Through a focus on community and survival we join rather than separate public opinion. Although the industry has a strong environmental argument for growth, it fails to build on this major asset. Nor does it acknowledge its failure to recognize the importance of the notion of global citizenship. The industry needs to recognize that the community's judgment of a healthy landscape is made using cognitive and intuitive processes, even though they are not always compatible!

In the British Isles the wind industry owes much of its present precarious position to its fragmented strategy for progress, one which does not adequately join local, national, and international interests. A universal language, one of compromise and negotiation, will help gain our shared interests of achieving a sustainable energy policy.

THE PATH TO FAILURE OR THE ROAD TO SUCCESS?

We need also to examine the cost of failure. We can turn down the right road by considering some fundamental flaws in the present combination of industry's cavalry charge for profit and expansion and government's lumbering attitude to global problem solving. First, collective removal of the public blinders to clean power will take more than painting lipstick on the gorilla. The industry must advocate public involvement in decision making and particularly communal ownership of wind farms. We must make site specificity a design requirement. Above all, the industry must consider the importance of public perception before, during, and after the development of a strategy and its implementation. One way to stimulate discussion and process is to raise questions, putting them in place simply to raise issues. This would all be part of a move to a more democratic and inclusive planning process which, I believe, would change public perceptions about wind energy.

THE ROLE OF THE ARTIST

Given my background in art, it is not surprising that I propose an increased role for the artist. To me there is a simple logic in incorporating an artist's skills and perceptions as a way of changing people's views of wind generation and the landscape. Given the strong aesthetic implications of wind power in the landscape, no group seems better prepared than artists to assist the public in expanding its perspective on the landscapes of power. Artists are equipped to help explore issues such as the quality of life, right and wrong, the beautiful and the ugly. What discipline better understands the anarchic nature of chaos? Artists suggest that chaos is as much a natural part of the environment as order, and, in fact, change and chaos are never far apart and are often interchangeable and always inevitable. What discipline visually explores and explains the intangible? The visual arts!

I have developed my beliefs from reading within wide disciplinary boundaries, especially writings which explore intersecting theories of culture and science, and that of art and sociology. Among others, John Urry's *Consuming Nature* and Simon Schama's *Landscape and Memory* come to mind.[1] These and many other works have convinced me of the inseparability of people from their place and the need for a holistic environmental aesthetic, one taking the public's views into consideration.

THE FUTURE LANDSCAPE SYMPOSIUM

English artists, architects, land managers, conservationists, lobbyists, environmentalists, academics, and those from sundry other disciplines joined thoughts at the 1996 Future Landscape: New Partnerships Symposium to discuss how to stimulate interdisciplinary approaches to planning and managing future landscapes.[2] Their deliberations profoundly affected my views on wind turbines in the landscape. Conference delegates agreed on one basic point: The British countryside is oversubscribed with multiple and often irreconcilable demands. The countryside is in conflict, and it is not just from wind turbines.

To illustrate, let me take you on an imaginary walk in my home county of Cumbria. It is clearly no longer the tranquil landscape of the poet William Wordsworth (Figures 2.2 and 2.3). While strolling in the hills one jumps deftly to avoid a deranged mountain biker, and steps right into the path of a vast crowd of nature ramblers hiking en masse. Then while veering away from this noisy crowd, you dodge the hail of gun shots from

FIGURE 2.2   A typical pastoral scene in the Lake District of Cumbria, one of England's cherished landscapes. (Courtesy Martin Pasqualetti.)

pheasant hunters, only to step into the pathway of red clad hunters and their fox hounds. While you observe the "keep out, danger" signs warning of mine subsidence, a high-speed train whistles through the valley, and you turn around in surprise near the "get off my land" sign just as a farmer threatens your dog. Just as you are thinking about the invisible radioactive pollution you suspect is coming from the nearby Sellafield nuclear works, you instinctively duck your head from low flying Tornadoes and Harrier jump jets. Finally, in your attempted escape to nature you find that the forest is no longer made of trees, but is now the whirling blades of wind turbines.

The landscape is no longer just a work place for a rural population. It is now a playground for the urban dweller, a highway for the tourist, and a resource site for energy. Although the problems associated with these demands are widely recognized, there is little agreement on how to resolve them! The 1996 symposium attempted to address that challenge. Most important, the use of wind energy as a case study initiated a partnership between contemporary artists and the industry. Discussions helped make it clear that artists have an important role to play in creating new ways of seeing wind turbines as icons for a sustainable future.

The symposium's central purpose was to define a new environmental aesthetic, one often at odds with Wordsworth's romantic paradigm. Initially it was the effect of his poetry on English thinking about landscape which

FIGURE 2.3   Some of the five Vestas 225-kW wind turbines at the former Haverigg air base, southwest Cumbria. The turbines are 27 meters (90 feet) in diameter and stand atop towers 30 meters (110 feet) tall. The turbines are located near the shore of the Irish Sea. Black Combe rises to the northwest and a prison borders the site to the east. Across Duddon Sands to the east, the turbines of Kirkby Moor are sometimes visible. The Lake District National Park is less than 40 kilometers (25 miles) to the north. (Courtesy Paul Gipe.)

led artists to use visual imagery to convey what became a completely romantic view of people and landscape. This view of the "noble peasant" is as potent today as it was then! Today, it is that inherited romantic gaze which is the public's expectation and one which the landscape must represent. No modernist world here; the landscape time machine is stuck in 17th-century Britain with happy serfs and caring landlords.

Alongside speeches and seminars at the symposium, artists and landscape representatives made presentations on existing collaborations, culminating in nine interdisciplinary creative workshops. Delegate and speaker groups were asked to respond to three real-life problems facing landscape and industry. These detailed "problematiques" were set by United Utilities, Cumbria County Council, and the British Wind Energy Association (BWEA). From the wind industry perspective, the conference represented growing support for wind farms. In particular, Lord Gowrie, Chair of the Arts Council of England and an ardent wind turbine supporter, suggested that "large constructs in the landscape can look wonderful. . . . Wind farms are part of the natural evolution of a working countryside." Coming from a member of the British aristocracy, the stridency of his call to modernize the romantic view of landscape was a sword slash to the mission of "Country Guardian" type organizations.

Most delegates supported the urgent need to alter present public perceptions of the British landscape to accommodate wind power. "It was a heady experience to be amongst people who were excited about what wind turbines could add to a landscape. . . ," said Colin Palmer, of Wind Prospect Ltd. Robert Lamb of Friends of Earth observed that no "planner, architect, ecologist or corporate sponsor would take the risks artists embrace, or share their urge to cross orthodox boundaries." Landscape architect Alison Parfitt noted that "accelerating change and increasing realization that we cannot go on living the way we do, suggest that the values of our society are up for question." Others emphasized that artists and local groups were already working in partnership. Finally, many emphasized the need to see landscape as a subject for interpretive debate, an approach that might make life more complicated for corporate managers.

THE ARTIST AS FACILITATOR

Perhaps the most important outcome of the symposium was the recognized potential role of the artist as facilitator and mediator in arriving at communal ideas of a project. It was clear from discussions that artists are best used as creative thinkers and doers, and that they work better if they are involved in a project right from its inception. John Kippin, one of the artists participating in the symposium, remarked: "It is important to stress the holistic approach . . . and to carefully consider the creative input of the artists as central to the conceptual process and not just as a decorative add-on." Another artist, Sasha Ward, remarked, ". . . Putting the different professions together at the earliest stage of the planning process and on the same level . . . enables us to concentrate on the process rather than the end product."

Artists can have a positive, inclusive effect, drawing out people who believe they are not part of the dominant culture, an especially important role to play when professions cross. At the conference the fear of having to talk to someone from another field was so great that at dinner people asked to be seated at tables representing their own discipline! My insistence on a variety of professions being at each dinner table resulted in such robust and meaningful conversation that the original complainants made a point of saying how much more they had learned from talking with people of different backgrounds and concerns!

LANDSCAPE AND COUNTRYSIDE

Many of the discussions in the symposium centered on how perceptions of rural landscapes vary, all too often it seems, irreconcilably. This was most evident in exchanges about landscape and countryside, a complex area with a plethora of interests and agendas significant to the success or failure of wind power development. Cultural critic Robert Hewison attempted to separate the meanings of landscape and countryside, suggesting that "Landscape is a concept, a mythical place where expectations have been set by history, through culture and art, but countryside is where you get your boots dirty!" Too often with landscape we miss this point; our reality is often different from what the media tell us it is, separate again from what we are inclined to believe as true, especially as we exist among the pressures of modern urban living.

Like beauty, interpretation of landscapes is subjective and selective. We carry ideals and notions which are combinations of media images and our own memories and associations. We subconsciously compare the reality (countryside) before us with these ideals (landscape). When they are not coincident, we feel a sense of loss and insecurity.

CHOCOLATE-BOX IMAGES

A nostalgic, "chocolate-box" image exists in Great Britain as a mental template for what constitutes "beautiful landscape." This image is a particularly strong trait among the British, a part of the Wordsworth paradigm that "countryside" equals "good." The chocolate-box images, which are reproduced upon nearly all boxes of such sweets, depict a series of mountains in the background, lush green valleys sweeping down to a small white cottage with a small family outside, and happy dogs and cattle "playing" in the sunshine. The reality is vastly different, and perhaps more in tune with this reality is novelist Fay Weldon's dire observation that "there's a dead man under every hedge, you know. He died of starvation, and his children too, because common land was enclosed, hedged, taken from him. ... The past is serious."

It is commonly observed that we Brits have become an urban people with rural longings, because of the small size of our island and the large size of our population. It is a trait which goes further back in Britain than in most other European countries, and surely helps explain the stridency of the objections to wind energy projects. This hankering after the country life was reinforced by the romantic movement, established by artists working in the 17th century, with its ennoblement of the rural existence that ignored and sanitized the realities of day-to-day life in the country-side. As Colin Palmer, of Wind Prospect Ltd. said at BWEA's 1997 Wind Energy Conference in Stirling, Scotland: "Selectively framed, perfectly composed representations of landscape, such as in Constable's paintings, shine in our consciousness as fixed visions of a lost, more idyllic and superior past. While there can be little doubt that we should cherish the landscape paintings of Constable the artist, it is questionable that we should cherish the expectations of landscape that these have fixed in our minds."

COUNTRYSIDE

If the landscape is an idealized image of the countryside, what is the true countryside? It is where people live, where they make a living, where they get their boots dirty. It is a place of conflicting economic and cultural demands and a place that is subject to rapid change. Patterns of land use, agricultural techniques, ownership structures, road systems, and waves of newcomers are all changing the countryside. Yet the outsiders do not want change. They wish to project their fixed, romantic images of landscape and scenery. They react negatively when the reality changes and diverges from these images. On the other hand, country people are more aware of being surrounded by a progressive past. They tend to see the land around them not as "17th-century static," but as a place where change is a constant.

In his classic book *The Making of the English Landscape*, W. G. Hoskins said, "You could write a book about every square inch of the ordnance survey map."[3] His notion is that landscape is one of constant change, that the text of the landscape has been rewritten from one generation to the next, where forest was replaced by pasture land, was hedged, mined, reclaimed, bombed, became a site for industry, used for houses, was cleared, and finally became a park for leisure time. The salient questions is, Which of these landscapes is more natural? Which more artificial? The answer is that they are all the same, a part of a changing countryside. It is the human view of landscape which is fixed.

In this decade of landscapes quickly changed by the addition of wind turbines, our attention is attracted because what is now seen is different from what we have grown to accept as "natural." It is in reality only another step in a series of ongoing evolutionary changes of a working countryside supporting human existence.

Of course one person's (working) countryside is another person's (pristine) landscape. Increasingly the expectation of change among those who live and work in rural areas clashes with notions imposed by city dwellers. For wind energy, conflict has come at a time when there is an increasing desire for something fixed, some point of reference, a turn toward the "permanence" of landscapes. Though the countryside has always changed, the rate of change is now accelerating. Wind energy development is only a part of a much wider set of quick changes that is challenging our society.

THE ARTIST'S ROLE IN WIND PROJECTS

There can be little doubt that in the past the term "landscape" was shaped largely by artists. Today, the expectations of a living environment are shaped more by cultural commentators such as journalists and public lobby groups. However, the artist can play a positive role in changing public perceptions. The battle for a new perception of wind farms needs creative proposals which are site specific, not solutions that are packaged and parachuted onto each proposed site. We must ask why the wind industry tends to reject the "mushroom" approach of slow growth from the bottom up. Perhaps too many engineers look for answers in equations rather than in the "soft" options of such intangibles as ownership, landscape memory, and quality of life. Having on numerous occasions approached the British wind industry to employ artists to facilitate change in the landscape, I find that they have been very reluctant to innovate. Perhaps this reluctance helps explain why they have had 75 percent of their site applications denied. Artists, working with the community, with planners, and with developers and the wind industry might change that percentage. They can create solutions which are acceptable to both sides.

In this new age of pluralism it is highly appropriate that the arts should begin to mix the economic needs of industry with the dreams and aspirations of the community. The artist as facilitator and communicator, and as *animateur* (someone who makes things happen), has grown to reflect the relative growth in the importance of communal and community interests. For example, "Art of Change," an artist group based in London, has evolved a practice which uses the community as the designer of landscape intervention, helping make it possible for the community to design the look of their environment. This can be a fruitful sort of involvement between industry and landscape. Thus, the artist can then exist as the intermediary, moving between the public bodies, the lobby groups, government and industry, all the while using the visual language we all understand.

THE CONSULTATIVE PROCESS

It is not surprising that the major wind companies fail in three-fourths of their proposals. Consultation and negotiation, in most cases, have been only superficial. In the main, wind companies use all the tact of gunboat diplomacy. If the industry is interested in changing the public's perception of wind energy landscapes, then it must be prepared to listen and compromise.

Such consultation does seem to reap rewards. Let us take the example of Peter Edwards, a private wind farm developer. Edwards included the community in his plans and his development of a small wind farm in Delabole, Cornwall, in picturesque southwestern England. Encouraging involvement and using education, Edwards helped locals toward a more positive tone. Another example is the community ownership of wind farms in west Scotland. There the World Wide Fund for Nature has helped the community understand its energy needs across a broad spectrum of alternative energy providers. For the most part, however, the result of a poor public relations strategy in the United Kingdom is appallingly evident. Proposed projects simply fail to get built because local planners fear community outcry expressed through the ballot box.

One of the principal barriers to greater development of wind power in British is the public's generally negative view of the technology. Generally this opinion is formed not by experience, but rather by ignorance, misinformation, prejudice, and fashion. It is manipulated by word of mouth, as well as by government/industry media and lobby groups. Individually, it is formed visually, aurally, intellectually.

Why has the wind power industry failed to win over a sympathetic public? The answer is that the industry underestimated how much the British people value landscape as a cultural resource, one central to their well-being. It failed at the marketing stage by neglecting to advance the idea that the *ownership* of wind power could and should be with the people. External or corporate ownership is almost guaranteed to bring out negative reactions in anyone. Now the majority of the public takes no part in the debate, but simply views the wind industry as aggressive and uncaring.

ESTABLISHING A SUSTAINABLE AESTHETIC FOR WIND FARMS

One way to improve public perception would be for the industry to agree on the components and functions of a sustainable landscape aesthetic. What is culturally acceptable? Can we chart how good or bad a new development would be for the public? To do so first we need some general ground rules. Such ground rules might include that (1) quality, beauty, and ugliness are reflections of personal taste and experience; (2) past models of landscape are often viewed romantically as being more desirable than the present; and (3) local views belong to local people, local history, and local memories.

We must also consider what views industry might bring to the table. Among the most important are that economic necessity is as valuable as spiritual meaning, that intervention and change are normal evolutionary characteristics in the landscape and countryside, and that "artificial" is a meaningless adjective, particularly when applied to the environment.

Since landscape is reactive to cultural, social, political, and economic factors, our aesthetic should include these elements as overall categories. But the fine tuning of a site-specific landscape analysis will require inclusion of more intangible factors to provide an adequate local aesthetic. Such factors will include art, nature, culture, history, socioeconomics, complementary associations, special characteristics, and tangible and intangible quotients. The very complexity of the task I am suggesting means that we cannot devise an environmental aesthetic which is acceptable to all people.

WHERE DO OUR AESTHETIC PROCESSES OPERATE?

Mental landscapes, that is, landscapes which exist only in our imaginations, often rely on associations with the past. As manifestations of dreams and desires, they are "props" we use to help us survive our present. These internal landscapes often control how we view the countryside and the possibility of wind turbines upon it. These "maps" are of primary importance to us and our survival. They are what locate us in a wider society. They make us who we are, distinct from our neighbor. Often our "internal" landscapes will stimulate a negative response to any suggested change. Why? Because such changes directly threaten our identity, ourselves. And, as previously noted, urban people feel they have the most to lose. In this regard, city dwellers are often seen as the worst troublemakers because wind turbines threaten their identity. In such cases, they insist that compensation must be paid: if not monetary, then it must be in new park lands.

THE IMMOVABLE ARGUMENTS OF NIMBYISM

No matter how clever and well-fashioned our mutual aesthetic is, how can it succeed when it comes up against the seeming contradiction of "yes, but not in my backyard"? A society might accept the need for balance, yet we find that beauty and order reign supreme. Each society strives to shape all to its sense of correctness or order. Thus, we know that in some situations landscape wilderness must exist as a means to define the concept of ordered landscape.

The 1995 Lusto Conference on Forest and Aesthetic in Finland questioned whether wilderness needs to exist as a reality, or whether the concept is enough.[4] Perhaps the English citizens with their eternal preoccupation with gardens, ordered and tidy landscapes, and planned parkland and rural estates provides the best clue to the importance of order and control to civilized values. This sense of order and control can be a formidable obstacle to change, as seen in the responses to wind turbines. Parochialism is the most prominent element in this power play, and although it is easy to understand, it is extremely hard to overcome. The NIMBY response is founded not on balance, but on a personal perception of balance, especially when it is expressed as a need to control one's immediate environment. In the NIMBY equation, beauty is in the eye of the beholder and its strength as a tool to maintain the status quo is reinforced through determined people, often at a cost which denies other local benefits. This kind of unbalanced viewpoint is essentially anarchic, one which is specific to any area where there is a confrontation between industry and the public. In any negotiations it would be wise to consider what trade-offs can be negotiated.

THE WIND INDUSTRY'S DILEMMA (AND CHALLENGE)

Through a series of events, the wind industry finds itself in a difficult position. After having failed to take fullest advantage of the initial positive public perceptions of their technology, it now finds a public attitude hardening against its interventions. Without radically changing the way it introduces wind farms to the landscape, it will continue to reinforce the view of an uncaring and profiteering industry. No longer can it afford to accept massive failure as an acceptable price to pay for limited success. For every gain the industry has suffered major losses.

A new approach is clearly needed. A minimum of £100,000 is spent by industry on each planning tribunal to decide on the right to build a wind farm. When only one of four is approved in the United Kingdom, the public is reassured of the rightness of its position. Ironically, the wind industry is paying for its abominable failure in public relations. If the same money had been spent on achieving shared or mutual ownership, community consultation, and/or compensation for landscape change, even a small gain would be money well spent in terms of wind farms built. Moreover, it would perhaps indicate a change in heart and strategy, thus winning back the approval of the public.

Today the wind industry is isolated. This is evident in its failure to win either specific approvals or the general battle of public opinion, despite what appears to be an uneven match between polluting energy sources and the cleaner environment wind energy promises to deliver. Obviously, the industry must change its approach to one that respects the value and place of public interests in negotiation and compromise. It must reject one-sided solutions in favor of paths toward understanding shared needs. It must make an investment in the way wind power is *marketed*. Finally, it must rely less on the gods of technology and efficiency and focus more on aesthetics and people. After all, the public is impressed by technological advances only briefly. They do not easily or quickly change their minds about interventions in the landscape!

CONCLUSIONS FOR A FUTURE STRATEGY

I have stressed here my belief in holistic solutions, that there are shared values in science, technology, and art. And, of course, we believe in the democratic process, trusting an informed public to make the right choice. Often that "right choice" will include wind turbines. But for this to happen, the wind industry must respect our cultural connection to the land, an attachment to the landscape that has been reaffirmed in the United Kingdom as a metaphor for national identity. We are, after all, tied to the land, its boundaries, and its climate. We feel a part of it. Many feel that the land actually shapes us. With these connections in mind, it should come as no surprise to the wind industry that when their plans challenge an existing landscape, people rise to vigorously defend their views.

In the early years of a new millennium, we are holding ever harder to the past. Whatever else might happen, the English devotion to landscape is not likely to disappear. As elsewhere, rural regions have become fantasy escapes for city dwellers. These urbanites treasure the countryside for what they see as its quiet, uncomplicated lifestyles and vistas. This tendency will increase. The massive population of centralized societies, as represented by London, has pressurized and distorted the value of landscape, shifting it away from the primacy of the rural community's need to survive.

In the countryside, views toward landscape are not inimical to industrial and agricultural change. However, the voice of the countryside has often been overwhelmed by a well-meaning, but largely dislocated urban population. For the new urban owners of the land, living in the countryside has also come to mean owning the view while working elsewhere. They do not share the values or concerns of the working farmers. Perhaps time will change attitudes. After all, sheep as gardeners in the Lake District are now accepted, in spite of their recent introduction. Rocky outcrops, the remains of ancient mining industry, are now considered national heritage sites. Electricity pylons have become organic. Perhaps in the decades to come wind farms will become acceptable heralds of a change to those advocating green power.

The wind industry must now contend with modifying a landscape which is seldom ever seen with a common view. Success will require collective consultation. If changing the landscape is critical to achieving environmentally clean energy, and if changing the landscape is a cultural issue, then it is time to use the language of art to influence public perception. Only then will we achieve the cultural compromises necessary for wind energy success.

NOTES AND REFERENCES

1. John Urry, *Consuming Places* (London: Routledge, 1995); and Simon Schama, *Landscape and Memory* (New York: Alfred Knopf, 1995).
2. "Future Landscapes: New Partnerships" (Sunderland, United Kingdom: Arctic Press, April 1997). Multidisciplinary environmental conference held in Windermere, November 1996.
3. W. G. Hoskins, *The Making of the English Landscape* (London: Penguin, 1999), 71. Ordnance survey maps are detailed topographic maps of the United Kingdom.
4. "Lusto Conference on Forest and Aesthetics," an international environmental conference on the philosophy of landscape aesthetics held in Lusto, Finland, June 1995.

3

THE WIND IN ONE'S SAILS: A PHILOSOPHY

GORDON G. BRITTAN, JR.

Placing wind turbines on the land can generate not only power but public opinions. Such reactions can diverge from one group to another, depending in large part upon one's philosophy. As a philosopher who uses a wind turbine to power his Montana ranch, Gordon Brittan is interested in historical and theoretical questions bearing on wind energy aesthetics. In this essay, he reflects on how opinions form, how we view technology, and how such feelings can affect the future of wind power development everywhere. He concludes that we do not resist wind turbines because they are uglier than other forms of energy production, but because they are characteristic of contemporary technology.

Let me begin with a fact. Separation of urban, rural-agricultural, and wilderness landscapes is essential to American attitudes and ways of life. I would suggest that this is true in many other countries as well. But conventional energy-generating technologies do not respect this separation. They dirty the air. They pollute the water. In a sense, they take the entire planet as their backyard. In my own state of Montana, largely undeveloped and remote, high mountain lakes show significant levels of acidity, likely due to air pollution from power plants. Fish populations below hydroelectric projects are altered dramatically. The mining of vast coal beds gives life in some parts of the state a strongly different character. In Montana the effects of conventional generating technologies are present everywhere. One would be naive to think that the effects of conventional energy-generating technologies can be kept out of any state if they cannot be barred from this northern, lightly inhabited state. If you can't hide such impacts in Montana, where can you hide them?

Copyright © 2002 by Academic Press.
All rights of reproduction in any form reserved.

Yet, even though wind generation is much more benign than conventional technologies, parochial opposition to it continues. In my home region, one of the windiest in the country, development efforts have been stymied to some extent by the fact that the impacts from wind energy are site-specific, whereas most of the impacts from conventional energy development are not suffered locally. Some of the opposition, however, is deeper, rooted in paradoxes associated with the wind generation of electricity in particular. Three of these paradoxes are familiar. Their resolution will be the focus of my discussion.

PARADOX #1

We have promoted wind generation of electricity to minimize the environmental impacts of energy production. Yet in its present wind-farm form of two- or three-bladed 100-Kw+ machines on 80- to 120-foot towers in extensive arrays, it creates a very noticeable impact; indeed, a number of major environmental organizations have been successful in limiting or thwarting its expansion. They take the position that conservation of electricity is the only acceptable alternative. This paradox of environmentalists fighting wind energy development exists because the impacts wind power is intended to mitigate are, for the most part, invisible, while the impacts of wind turbines are clear and unavoidable.

PARADOX #2

In its present form, wind power is possible only in rural or otherwise relatively undeveloped areas where, in consequence, the visual impact it makes is greatest. As these rural areas progressively shrink, the desire to preserve them in a pretechnological condition becomes greater.[1]

PARADOX #3

Our present methods of generating electricity from the wind will become generally acceptable through familiarization, but it will take at least a decade. Most likely they will be "online" *after* the time when wind power is most needed to mitigate the effects of the more technologically sophisticated "hard energy path."[2] Particularly in the western United States, we now face a large capacity shortfall and hence an immediate need to install new generation facilities or reduce consumption.

WAYS OF RESOLUTION

We need to resolve these paradoxes, though I don't think it will be possible to do so without prying open larger questions concerning the nature of contemporary technology and the organization of our social lives. That is, I don't think that currently suggested solutions to these paradoxes will be successful. In my view, the instinctive opposition some express toward wind turbines is entirely reasonable, given the present manner of wind development in California and elsewhere. If wind power is going to grow in importance, we must change our ways, that is, we must reorient the ways in which we think about wind energy and not merely try to mitigate its present impacts. Here are four such ways, all of which have in common that they leave both the technology and the social context untouched. I wish to discuss (1) the siting problem, (2) the people problem, (3) the perceptual problem, and (4) the marketing problem.

SITING

I believe that the masses of wind machines seen at places such as Altamont Pass and San Gorgonio Pass in California do not simply transform the landscape, they threaten *us* as well. However, the visual impact can be very much diminished, if not eliminated entirely, by breaking the arrays into clusters of approximately 10–15 machines. Another solution would be to space the wind turbines across the country-side one at a time, an arrangement both familiar and acceptable in most parts of the world.

One difficulty with this more distributed arrangement is that when a wind turbine breaks down, it is less easily repaired. Perhaps my personal experience is germane. When we installed a 65-kW Danish machine on our ranch in 1985, major repairs always required that we fly in an engineer from Copenhagen. As a result, there were long periods during which it did not operate. When we asked the engineer to explain what he was doing, in the hope that the next time around we might be able to do the same job ourselves, he replied simply, "Much too complicated." We have vastly simplified the circuitry since then. Still, it is at the limit of my own powers to make major repairs. In the meantime the turbines have become still more complex. I saw the original prototype of the U.S. Windpower 33M-VS turbine spread out across the floor at the company's Tracy, California, headquarters. Only a specialist could begin to understand it. But specialists cannot be hired to take care of one or even a small number of machines. This is one reason why large numbers of machines have been clustered.[3]

PEOPLE

Evidence suggests that people themselves are at the core of the frustrated growth of wind energy. Wind turbines are located, as in California, too close to population centers or along heavily traveled highways. We need to move the machines far away from these centers and highways, on isolated mountain ridges or out to sea. The farther away from people, the fewer the complaints and the more electricity can be produced.

This, of course, is an idealized suggestion. Its drawback is that such removal is often not a financial option. It has been estimated that almost 90 percent of the wind energy available in the Northwest area of the United States is located on the Blackfeet Indian Reservation in northern Montana. Why is there no development of wind power there? The three most apparent reasons are inadequate transmission lines across the Reservation, legal difficulties in securing a power corridor, and prohibitive costs for line construction.

PERCEPTION

When we perceive a wind farm, we balance benefits and costs. Every paradox can be resolved when its various elements are weighed and the trade-offs are made clear. Whatever disadvantages are to be associated with wind energy, they are more than offset by its benefits. Visual appearance must give way to environmental reality, which is that wind power, compared fully with other options, is the most benign energy source we have.[4]

However, the idea that we must accept unsightly wind turbines in the interests of the greater environmental good is unappealing. The grudging "You must eat your spinach" directive works only slightly better with children than with adults, particularly since there seems to be a clear alternative, namely to put the turbines anywhere else but "my" view. No one wants to have their backyard become a sacrifice area, regardless of the benefits for everyone else.[5] This factor may be at work in California where, despite large generating-capacity shortfalls, politicians (while giving some lip service to alternative sources of energy, including wind) are not generally calling for the expansion of existing wind farms.

MARKETING

Some wind energy advocates believe the problem of public opposition is a "marketing" problem. Using advertising and positive images, promoters believe that they can condition us to see large wind turbines in expansive arrays as beautiful.[6] If beauty is, after all, in the eye of the beholder, they believe they can alter what the eye perceives. At least in theory, given the increased familiarity the public has gained through the regular use of wind turbines, their appearance in films and commercials, and the cleansing of inefficient and nonperforming machines from the landscape, and given the right sort of advertising and promotional campaign, it seems quite likely that more people will come to regard them as beautiful and opposition to them will slacken.[7]

This fourth suggested approach to public opposition to the aesthetic problem provided the main topic of discussion among the authors of this book at the Rockefeller Foundation's conference center in Bellagio, Italy. It will require time, however, to consider it adequately. To do so we have to take up some general historical and theoretical questions and leave the topic of wind turbines temporarily. We need, first of all, to be clear about how people respond to landscapes. We need to realize that such response is not simply a question of manipulation and control but, more fundamentally, it is more a matter of knowledge than belief.

THE NOTIONS OF NATURAL BEAUTY AND SCENERY

Unlikely as it may seem, the notion of natural beauty, with its general appreciation of certain sorts of landscapes, seems to have developed in Western culture in the 18th century. At the same time, not coincidentally, aesthetics emerged as a separate philosophical discipline, as did our modern classification of the fine arts. Eighteenth-century taste in music, drama, and painting have variously given way, but the notion of natural beauty and of what constitutes scenery continues to dominate much of our thinking today.[8]

The main lines of thinking about these two themes are very familiar. Ideal landscapes are considered to be balanced, not only in terms of their form or composition, but with regard to their content as well. Human elements are well integrated with natural elements. Everything in the landscape is to scale. The whole has the appearance not of a French garden, but of an English park.[9] However informal, even wild, it might appear, order is everywhere. And it is, in the usual term, picturesque, and therefore attractive.

How do we think about scenery? Let me offer three comments. First, our perception of landscape as scenery is not derived from direct contemplation of nature, but rather from the tradition of 16th- and 17th-century European landscape painting.[10] Indeed, photographs of natural landscapes, stripped of all classical allusions, still tend to resemble in their overall "look" (color, composition, and use of light) the works of Claude Lorrain and Nicolas Poussin. It follows as well that our standard of natural beauty is primarily visual in character.

Second, the 18th-century notion of scenery is not simply conventional. By this I mean that it was also informed by the scientific discoveries of Galileo, Newton, and others. These discoveries allowed the landscape painters to see, and consequently to appreciate, a certain mathematical pattern and regular order in the landscape, and to render it using the laws of perspective and a deeper understanding of color. However disorderly it might at first appear, we know that there is design in nature, a design which particular arrangements of form and color can reveal. We see with the mind what is not immediately apparent to the eye, the work of an intelligent and powerful being.

Finally, and to my point here, it is difficult if not impossible to reconcile contemporary wind turbines on the landscape with this 18th-century ideal.[11] They dominate rather than harmonize. They upset rather than balance. They are not to scale.[12] There is no place for them in the "park." It follows, therefore, that they are "ugly."

The question is this: is there another standard of natural beauty? Is there a standard that is informed by scientific knowledge not available to the 18th century, that is not principally visual in character, and that is not similarly violated, which allows us to go beyond the picturesque and the pretty? If so, then we should move to disclose it, and in this way make an aesthetic case for wind energy landscapes.

ALDO LEOPOLD

In addressing this question, there are a variety of alternatives we might consider. But for my purposes, and because it is so closely aligned with what might be called an "environmental aesthetic," I want to focus on the view of the natural philosopher Aldo Leopold.[13] More than any other American since Thoreau, he tried to show us how to reap from the land "the [a]esthetic harvest it is capable, under science, of contributing to culture."[14] The science he had in mind was not physics. It was biology.

In his essay "Marshland Elegy," Leopold set out to instruct us in the beauty of the swampy sort of country traditionally either "reclaimed" through draining or ignored as wasteland, similar to the Wesermarch district of Lower Saxony that Christoph Schwahn discusses elsewhere in this volume. Leopold's account centered on a native marsh inhabitant, the sandhill crane.[15] The crane itself is beautiful not so much by virtue of its appearance as because of its evolutionary history, the way in which it symbolizes our very ancient and untamable past.[16] But the crane is also "interlocked in one humming community of cooperations and competi-tions, one biota," with all inhabitants of the marsh which thus inevitably share its beauty.[17] Leopold's aesthetic makes a place for balance in terms of the detailed way in which the activities of plants and animals play off against and compensate one another rather than the arrangement of masses or the equilibrium of physical forces. The great biologist D'Arcy Thomp-son once commented to the effect: "Things are what they are because, being what they are, they got to stay that way."[18] Leopold adds that in virtue of the fact that plant and animal communities have developed over long periods of time so as to maintain themselves more or less intact, they are beautiful.[19]

I offer here two comments about this aesthetic. First, it is not primarily visual in character. What makes a natural scene beautiful is not how it looks, but the way in which it expresses an underlying harmony which is itself the product of a long evolutionary history. Indeed, Leopold insists that to discover natural beauty we have to go beneath the appearance of things[20]:

> Ecological science has wrought a change in the mental eye. Daniel Boone's aesthetic reaction, for example, depended not only on what he saw, but on the quality of the mental eye with which he saw it. It has disclosed origins and functions for what to Boone were only facts. It has disclosed mechanisms for what to Boone were only attributes. We have no yardstick to measure this change, but we may safely say that, as compared with the competent ecologist of the present day, Boone saw only the surface of things. The incredible intricacies of the plant and animal community — the intrinsic beauty of an organism called America then in the full bloom of her maidenhood — were invisible and incomprehensible to Daniel Boone.

The merely picturesque is trivial for the same reason.

Second, Leopold's aesthetic is pluralistic. By this I mean that any sort of natural object or system of objects is beautiful insofar as it is what it is because, being what it is, it "got to stay that way." In both of these respects, Leopold's aesthetic is revolutionary. It opens us up to appreciate, at a deeper level, landscapes that were beneath the notice of the 18th-century perspective. It leads our age to preserve and to cherish what was formerly regarded as unattractive, chaotic wilderness.

WIND TURBINES AS BEAUTIFUL

But Leopold's view does not open us up to the appreciation of every-thing, certainly not to everything which is conventionally pretty. In particular, I don't think that it will help us make a case for contemporary wind turbines. This is not because any human presence or artifact in the landscape necessarily unbalances it; we, too, have evolved over long periods of time, as have some of our artifacts, to the point where at least in certain communities some sort of equilibrium condition has been reached. Rather, it is because the turbines are a new and exotic species. As Baird Callicott puts it, "From the point of view of the land aesthetic, the attractive purple flower of centauria or the vivid orange of hawkweed might actually spoil rather than enhance a field of (otherwise) native grasses and forbes. Leopold writes lovingly of draba, pasqueflowers, sylphium, and many other pretty and not-so-pretty native plants, but with undisguised contempt for peonies, cheat grass, foxtail, and other European imports and stow-aways."[21] Some new and exotic species threaten to upset the at least temporary equilibrium of the biotic commu-nities into which they are introduced, and for that very reason must be resisted. But I take it that on Leopold's aesthetic scale, even nonthreaten-ing new and exotic species cannot be beautiful. It follows that wind turbines, however little they otherwise disrupt the biological integrity of particular landscapes, cannot be beautiful.[22] Simply put, they lack the right sort of history and they are not organic.

This point needs elaboration. No species is alien per se, but only with respect to particular environments. It is a matter of context. But in many of the environments in which wind turbines have been introduced, they are unacceptable fantasy creatures. At some future time, they will no longer be fantastic (having evolved along with us and other creatures, in particular biota). They will have become acceptable. Alas, in the context of the third of our original paradoxes, by that time it will be too late. Ironically, 50 years from now their coming to be regarded as beautiful will be a function of their having become useless.

Of course, it is impossible now to put them into a biological perspec-tive. It is safe to say that they are a new human adaptation, introduced at a time when conventional energy-generating technologies are no longer adaptive, a half-conscious attempt to bring ourselves once again into some sort of stable relationship with our environment. But it is not enough to say this. Wind turbines must also have the "right sort of history," in fact they must *prove* their adaptive character, in order to qualify as beautiful. There is no way to say in advance whether they fit into particular biotic communities; they are like so many other human artifacts and activities for which we don't as yet have an adequate evolutionary perspective and cannot therefore call beautiful.[23]

TO GO BEYOND APPEARANCE

We can still follow Leopold's lead. That is, we can ask ourselves: Are there other than classically physical or biological ways in which we can go beneath the mere (that is to say, conventionally uncomfortable) appearance of a wind turbine array? Can we appreciate some sort of deeper complexity and equilibrium in the way that Leopold urges us to go beyond the conventionally uncomfortable appearance of a marsh to the appreciation of a deeper ecological beauty? I think the answer to these questions is "no." This is not because of design features intrinsic to wind turbines. Rather, it is because of certain general features that they share with much contemporary technology. These are the features which, at least in part, stimulate much of the present resistance to wind energy.

Let me put the point this way. Wind energy (in its most recent embodiment) was introduced in terms of a "trade-off," one benign technology being substituted for malignant technologies. But aside from their benign and malignant features, these technologies share the same general design characteristics. And they were and are imposed, grouped, and owned in very much the same sort of way, a point to which I will return later. In my view, the resistance to wind turbines is not because they are uglier than other forms of energy production, but because they are characteristic of contemporary technology, magnified by their large size, the extensive arrays into which they are placed, and the relative barrenness of their surroundings.

THINGS AND DEVICES

In order to better understand the implications of wind power, we need to become clearer about the character of contemporary technology. No one has done more to clarify it than the philosopher Albert Borgmann.[24] In summary, Borgmann makes a distinction between "devices" (those characteristic inventions of our age, among which a pocket calculator, a CD sound system, or a jet airplane might be taken as exemplary) and what Martin Heidegger calls "things" (not only natural objects, but such human artifacts as the traditional windmills of Holland).[25] The pattern of contemporary technology is the device paradigm, which is to say that technology now has to do more with devices than things.

Things engage us, an engagement both of body and mind, an engagement that demands skill. A device, by contrast, makes no demands on skill, and therefore disengages and disburdens us. It is defined in terms of its function. Usually it is a means to procure some end. Since the end may be obtained in a variety of ways — that is to say, since a variety of devices are functionally equivalent — a device has no intrinsic features. But a device also conceals, and in the process disengages. It obtains its ends in ways literally hidden from view. The more advanced the device, the more hidden from view it is. Moreover, concealment and disburdening go hand in hand. The concealment of the machinery ensures that it makes no demands on our faculties. The device is also socially disburdening in that it is completely anonymous.

To make the analysis of devices more precise, an objection to it should be considered. Borgmann asks, "Is not . . . the concealment of the machinery and the lack of engagement with our world, due to widespread scientific, economic, and technical illiteracy?"[26] He is explaining why, at least in principle, we cannot go inside contemporary devices, or break through their apparent concealments. Why should we not promote electrical engineering, for example, as a general course of study, and in the process come to know if not to love contemporary technology?

But Borgmann initially answers this objection along three main lines. First, many devices (such as the pocket calculator) are in principle irreparable; they are designed to be thrown away when they fail. In this case, there is no point in going into the device. Second, many devices (such as the CD sound system) are in principle carefree; they are designed so as not to need repair. In this case, it is not necessary to go into such devices. Third, other devices (such as the jet plane) are in fact so complex that it is not really feasible for anyone but a team of experts to go into them, something that is increasingly also true of older technologies, such as automobiles, where fixing becomes tantamount to replacing.

But Borgmann contends that even if technical education made much of the machinery of devices perspicuous, two differences between devices and things would remain. Our engagement with devices would remain "entirely cerebral" since they resist "appropriation through care, repair, the exercise of skill, and bodily engagement." Moreover, the machinery of a device is anonymous. It does not express its creator, "it does not reveal a region and its particular orientation within nature and culture." On both accounts, devices remain unfamiliar, distant and distancing.

THE BLACK BOX

We could summarize Borgmann's position by referring to the familiar theoretical notion of a "black box." In a black box, commodity-producing machinery is concealed insofar as it is both hidden from view or shielded (literally) and conceptually opaque or incomprehensible (figuratively). Moreover, just those properties that Borgmann attributes to devices can equally be attributed to black boxes. It is not possible to get inside them, since they are both sealed and opaque. Nor is it necessary to get inside them, since in principle it is always possible to replace the three-termed function of input, black box, and output, with a two-termed function which links input to output directly. But given all of this, then there is no deeper way in which devices can be appreciated, no informed perspective from which they are beautiful.

WIND TURBINES ARE DEVICES

Now I want to make a very controversial claim. Wind turbines are for most of us not things but merely devices. There is therefore no way to go beyond their conventionally uncomfortable appearance to the discovery of a latent mechanical or organic beauty.[27] Thinking for a moment reveals that except for the blades, virtually everything is shielded (including the towers of many turbines), hidden from view behind the same sort of stainless steel that contains many electronic devices. Moreover, the machinery is distant from anyone save the mechanic.

The lack of disclosure goes together with the fact that wind turbines are merely producers of a commodity, electrical energy, and interchangeable in this respect with any other technology that produces the same commodity at least as cheaply and effectively. The only important differences between wind turbines and other energy-generating technologies are not intrinsic to what might be called their design philosophies. In other words, although they differ with respect to their inputs (i.e., fuels) and with respect to their environmental impacts, the same sort of functional description can be given a fossil fuel plant. There is but a single standard on which to evaluate wind turbines. It should not be wondered at that they are, with only small modifications between them, so uniform.

Many astute commentators would seem to disagree with this judgment. Thus, for example, Robert Thayer in *Gray World, Green Heart*:[28]

> With wind energy plants, "what you see is what you get." When the wind blows, turbines spin and electricity is generated. When the wind doesn't blow, the turbines are idle. This rather direct expression of function serves to reinforce wind energy's sense of landscape appropriateness, clarity, and comprehensibility. In the long run, wind energy will contribute to a unique sense of place.

However, Thayer reinforces the device-like character of wind turbines. Only their function is transparent, wind in — electrical energy out. The black box where all the processing takes place remains unopened.[29] This is roughly the same kind of comprehensibility that is involved when we note the correlation between punching numbers into our pocket calculators and seeing the result as a digital readout.

There are two more things to be said about Thayer's position. One is that nothing can be appropriate to landscape per se; everything depends on the type of object and the type of landscape, at least if we think of landscapes, following Leopold's definition, in biological terms. It is a matter of context. But as is typical of devices generally, contemporary wind turbines are context-free; they do not relate in any specific way to the area in which they are placed (typically by an outsider). In particular, Leopold insists on the fact that the appropriateness of objects in landscapes has to do with their respective histories, the ways in which they evolved, or failed to evolve, together.

But contemporary wind turbines have only a very brief history, and in terms of their basic design parameters — low solidity, high rpm, low torque — differ importantly from the windmills whose history goes back at least 1200 years. If wind turbines have any sort of context, it is by way of their blades and the development of airplanes. However, it is difficult to see how airplanes fit as appropriate objects or symbols into a windswept landscape. Of course, in the long run wind turbines will contribute to a sense of place, but not simply in virtue of having been installed somewhere in massive arrays. They will first have to acquire a history.[30]

AN ALIEN ARCHITECTURAL ARRIVAL

It is interesting to note in this respect how unlike other architectural arrivals contemporary wind turbines are. Different styles of architecture developed in different parts of the world in response to local environmental conditions and the spiritual and philosophical patterns of the local culture. As a result, they create a context, or in Heidegger's wonderfully dark expression, they "gather." But there is nothing local about contemporary wind turbines. They are ubiquitously and anonymously the same, alien objects impressed on a region but in no deep way connected to it. They have nothing to say to us, nothing to express; they conceal rather than reveal. The sense of place that they might eventually engender cannot therefore be unique.[31]

The other comment I want to make regarding Thayer's position is that wind turbines are quintessential devices. They preclude engagement. The primary way in which the vast majority of people can engage with them is visually. They cannot climb over and around them. They cannot get inside them. They cannot tinker with them.[32] In most places, particularly in the United States, they cannot even get close to them. There is no larger (nontrivial) physical or biological way in which they can be appropriated or their beauty grasped. Should we be surprised that most people find them visually objectionable? Perhaps they might be willing to countenance their existence, but only as the lesser of evils.

So, in summary, there is not an immediately available aesthetic norm on which wind turbines are beautiful. Nor is there an immediately available and adequate conception of landscape which they fit into.

AN ARGUMENT FOR LOCAL CONTROL

I said earlier that the sheer complexity of contemporary wind turbines demands that they be grouped in rather large arrays, so that installation, maintenance, and repair costs can be minimized.[33] This entails, in turn, that they be owned and operated by large companies. Like other energy-generating technologies, their immediate context is industrial. But this fact is problematic for a variety of reasons. To begin with, the sheer size of the standard array is visually imposing and objectionable. Typically, they so completely dominate the horizon that it is difficult to integrate them with their landscape, even in a rather distant perspective. Furthermore, the fact that these arrays are owned and operated by large companies, whose bankers and boards of directors live and work far away from the site, diminishes any sense of local connection and, more important, of local responsibility and control.[34] Those who make the decisions regarding wind farms are not the same people who must live with them on a daily basis. As a country we have been slow to learn this, but those on the ground, who have a sense of the bounds of both tradition and environment, in general make the best land use decisions. E. F. Schumacher put it accurately when he wrote: "It is obvious that men organized in small units will take better care of their land or other natural resources than anonymous companies or megalomanic governments which pretend to themselves that the whole universe is their legitimate quarry."[35]

Two points must be emphasized in this regard. One is that wind energy can grow out of local communities, which means that the turbines are sited, owned, and operated by local residents, or they can be imposed from outside, so to speak. In the former case, it begins to have that sort of "organic" connection to the whole which characterizes Leopold's notion of natural beauty. In the same way, it begins to express the life of the people who live there, as something they have freely chosen.[36] The best expression of this model is, of course, Denmark. The question of local control, as with individual comprehension, is thus closely tied to aesthetic apprehension. What we cannot understand or control might be sublime, but it can never, for the same reason, be beautiful. There is always and necessarily the question of scale.

The other point to emphasize is that local communities tend to have some sort of biological basis.[37] They are defined at least in part by the plant and animal life of the region, the kind and quality of the soil, the available rainfall and adjacent watersheds. It is important to realize that communities are characterized not only by mutual trust and a willingness to sacrifice for the common good, but also in terms of place and of history.

THE IMPORTANCE OF PLACE AND HISTORY

Although place is often identified with an individual terrain and a particular watershed, it could just as well be identified with a "windshed." In Montana, the winds come in the middle of winter when we most need them, raising temperatures and blowing the snow off the ground and providing electrical energy to heat homes. We call them *chinooks*. They are part of our lives, in the same way that the *mistral* is part of the life of the Midi, the *bise* of the Lavaux, and the *Föhn* of the Schwarzwald. There is even a playful little wind which swirls around the church of the Gesu in Rome. To treat them as no more than another energy source, a standing reserve as Heidegger would put it, is to disconnect them from the ways in which they have helped determine the character of local plant, animal, and human communities, and in the process to rob them of their individuality and their beauty. By the same token, unique windsheds need to be connected in specific (not simply functional) ways to wind turbines if the latter, in turn, are to share in this beauty.

TWO DIRECTIONS

I don't want to overemphasize the communitarian and the bioregional perspectives, although they should be important elements in our thinking.[38] The point is that these perspectives allow us to establish an aesthetic which is not simply conventional or visual, and on which both wind and the machines that capture its energy are beautiful.

There are in my view two directions to take. Taking one, we should encourage small and simple machines which can be locally owned and operated, without the intervention of a specialized engineer. Second, we must incorporate machines that have a history,[39] that supply a context, that are sensitive to their sites, and that as a result integrate with at least some landscapes and hence with the communities that have grown up on them. Again to quote Schumacher, whose thinking has shaped my own: "What is it that we really require from the scientists and technologists? I should answer: We need methods and equipment which are cheap enough so that they are accessible to everyone; suitable for small-scale application; compatible with man's need for creativity."[40]

A MODEST PROPOSAL

What, then, do I propose? First, we might consider a very different sort of wind turbine. A group of us has been working on its development for the past 20 years, although in fact the idea can be traced back to Crete where thousands of such windmills have been spinning for generations on the Lesithi Plain. In a very schematic way, let me draw your attention to its main features. The main design parameters are traditional — high solidity, low rpm, high torque. The rotor consists of sails, furled when the wind blows hard, unfurled when it does not. The machinery is exposed and thoroughly accessible, clear and comprehensible. All of it can be repaired by someone with a rudimentary knowledge of electronics and mechanics, and with the sort of tools used to fix farm machinery. It can be owned and operated by a person of modest means. It is situated at ground level and does not require a crane for either its installation or its repair.[41] It is a

FIGURE 3.1   Soft-foil turbine illuminated against a dark sky near Livingston, Montana. (Courtesy Gordon Brittan.)

downwind machine and tracks easily and freely. It is, therefore, a thing and not a device.[42] All of Borgmann's criteria are satisfied (Figure 3.1).

This experimental machine points to the fact that there is a need for creative design: Designs which are efficient, yet more in tune with their environment and what I have described as "things." Creative thinking should be stimulated through public and private capital. We should not assume that the three-bladed Danish turbine is the final and only option for wind gathering on the landscape.

The need for creativity is even more pronounced as we enter into this new phase in which electrical energy is being deregulated and decentralized, just the sort of development that Schumacher and others had in mind. It will, I believe, be more and more possible for owners of small numbers of wind turbines (and of the cooperatives into which I see them forming) to put their power on the grid, particularly since wind-generated electricity will never amount to more than 10 percent of the total.[43]

There are, of course, a number of problems with this scenario, but I think that groups of relatively small machines, working together, will ultimately prove to be more efficient, as well as more beautiful, than a single very large machine, in the same way that a number of smaller processors, operating in parallel, have supplanted, in many respects, very large mainframe computers.[44]

A PLURALISTIC APPROACH TOWARD CREATIVE DESIGNS

Finally, and again following Leopold's lead, I want to urge a pluralistic approach. If we pay attention to the beauty of landscapes, then we must conclude that certain kinds of turbines will fit some of these landscapes better than others. Certainly, I have tried to make a case for our own soft-foil turbine. However, there are other designs, some of them not yet imagined, which will no doubt fit their own landscapes. Design engineers must think creatively. Unfortunately, governments and utilities have not encouraged such creativity, content to focus on fine-tuning the three-bladed turbine, rather than a more aesthetically acceptable machine.

Along the same lines, too much effort, I would argue, has been devoted to making this same design palatable to the general public. Most of the papers in this collection take this as their general theme as well. I think we need to move in the other direction, by opening up the design and aesthetic question, a question which, as I've tried to indicate, cannot very well be separated from the character of contemporary technology or the nature of biological and human communities.

It is not enough to try to sell wind energy. On this basis, everyone buys it if only the machines are placed in someone else's backyard. To successfully promote wind power, we must develop instead comprehensible, efficient, site-sensitive, locally owned and controlled designs: turbines which we can relate to and have close by. We must also, as Karin Hammarlund and others have insisted, provide people with some sort of choice beyond a simple "yes" or "no." For in the final analysis, aesthetic questions begin to merge with moral ones. When we have learned that, we will indeed put wind in our sails.[45]

NOTES AND REFERENCES

1. Not simply because turbine arrays require a great deal of room, but also because the windiest areas of the world tend naturally to be less settled.
2. At which point they will undoubtedly become cherished reminders of a precious stage in our nation's history to be preserved.
3. For this and other reasons (among them the price of land), wind turbines are becoming larger and larger. The failure of multimegawatt machines in the 1960s and early 1970s

prompted the development of much smaller units; the pendulum is now swinging in the other direction and 700- to 900-kW turbines are soon to be the industry standard. Whether these behemoths, whose rotors are now 150 feet (50 meters) in diameter, will prove to be more visually acceptable than comparable arrays of smaller machines remains to be seen. It should be noted that U.S. Windpower (Kenetech) is now bankrupt, partly because of the complexity and frequent failures of the 33M-VS turbine.

4. Thus Robert Thayer, in his important book *Gray World, Green Heart: Technology, Nature, and the Sustainable Landscape* (New York: John Wiley and Sons, 1994), p. 127: "Gradually in the last few years, the public has begun to recognize that the benefits of such a relatively benign and renewable energy source far outweigh the impacts, most of which are visual and not geophysical or ecological in nature."

5. It is often said that the visual impact of wind turbines is a "small price to pay" for clean energy, when in fact it is unclear to many people whether any price has to be paid. It is the lesser of evils for a society which doesn't want to recognize any evils.

6. There are, of course, those who *already* think they are lovely. Robert Righter, whose *Wind Energy in America* (Norman, Oklahoma: University of Oklahoma Press, 1996) is the standard reference work, quotes several of them. For example, on page 249, a freelance writer reports that "they are not ugly, these wind turbines, bristling on the green crest of the Alameda hillside like a sparse mohawk . . ." But the simile employed here would only confirm for the majority their view that they *are* ugly.

7. Perhaps including books such as this.

8. Thus Frode Birk Nielsen in his contribution to this collection: "The goal is to establish a beautiful and narrative composition in relation to water or land surfaces, a visual balance between elements in the landscape created by man and nature, a whole."

9. Stourhead, the famous Wiltshire garden in England, devised by Henry Hoare, might be taken as paradigm.

10. See Christopher Hussey, *The Picturesque: Studies in a Point of View* (New York: G. P. Putnam's Sons, 1927): 1–2.

11. Robert Righter includes a photograph in *Wind Energy in America* (page 264) that he captions "A harmonious, indeed aesthetic, image of wind generators . . . at Altamont Pass . . . ." The photograph is striking for at least two reasons. One is that it is straight out of Lorrain and Poussin, by way of Ruysdael and Hobbema: very low horizon, (nimbus) cloud-filled sky, dramatic diagonal sweeping across while receding into the landscape, large shadow-casting boulders in the foreground. Only the first turbine might be said to dominate its context; the others (the nacelle of the third is already level with the low horizon) simply trail off into the distance. With a very different subject matter, the place might well be Calvary. The other reason the photograph is striking is related to the first: it is that the beauty of the image is only indirectly related to its subject matter, in the same sort of way that Walker Evans' images of distressing Southern poverty have their own transcendent beauty.

12. The 18th-century conception of natural beauty included objects not to scale. They were *sublime* and not strictly *beautiful*. But except by way of an occasional metaphorical extension, only natural objects, Mont Blanc or the Rheinfall say, could be sublime; out-of-scale human artifacts such as 250-foot (75-meter) towers arising without context from a windswept desert were merely grotesque.

13. Following J. Baird Callicott's essay, "The Land Aesthetic," in *A Companion to the Sand County Almanac* (Madison: University of Wisconsin Press, 1987): 157–185.

14. Aldo Leopold, *A Sand County Almanac, and Sketches Here and There* (New York: Oxford University Press, 1949): viii.

15. *A Sand County Almanac.*

16. Although it is, in fact, a very striking bird.

17. Aldo Leopold, *Round River, From the Journals of Aldo Leopold,* edited by Luna Leopold (New York: Oxford University Press, 1953): 148.

18. D'Arcy Thompson, *On Growth and Form* (Cambridge, England: Cambridge University Press, 1917).

19. Interestingly, Leopold's paradigm, the marsh, also figures in Christoph Schwahn's paper elsewhere in this volume. He said, "The marshes of Friesland along the North Sea Coast are extremely flat and could be called monotonous by someone who is not used to this special kind of landscape." But Schwahn goes on to indicate that mere familiarity is not at stake, "For myself, without a systematic landscape analysis I would have been lost in trying to localize differences in landscape structures. It was quite interesting for us to find out that the landscape units which were a result of our analysis corresponded with different epochs of marsh formation." Of course, what Schwahn sees as coincidental, Leopold takes as necessary, and in the process draws our attention (by instructing our perception) to the beauty of the marsh.

20. *Round River, From the Journals of Aldo Leopold,* 177.

21. *Companion to a Sand County Almanac,* 162–163.

22. The point must be emphasized. The reaction to wind turbines (and to other similarly scaled technologies) is in part a function of the relative fragility of the environments into which they are introduced. For wind turbines, as presently arrayed, must be introduced into relatively unpopulated areas and the factors which allow for human settlement in numbers are the same factors which allow a biological region to be relatively resilient. The fear, however unfounded it is, that wind turbines will disturb the San Gorgonio Pass has roots deeper than a desire to keep the view unspoiled.

23. One way to put the aesthetic issue is with respect to weeds. Laurie Short and others in this volume who take the subjectivist line think that weeds are simply plants that we human beings happen not to value. I agree with Leopold, rather, that weeds are such because (relative to particular environments) they don't fit in, they are invaders and (noxious) increasers, opportunistic outsiders who do not know, still less respect, their own place in the ecosystem. Some people think of wind turbines in the same sort of way as weeds, spreading beyond control and in some sense taking over the areas in which they are placed and overwhelming their competitors. But this judgment is at least in part premature; wind turbines (with all appropriate qualifications) are newly arrived. It will take some time to see whether they will fit in.

24. Albert Borgmann, *Technology and the Character of Contemporary Life* (Chicago: University of Chicago Press, 1984, reprinted 1987).

25. See Martin Heidegger's essay, "The Thing," in *Poetry, Language, Thought,* translations and introductions by Albert Hofstadter (New York: Harper & Row, 1971): 163–182.

26. Ibid., p. 47.

27. Presumably there is a group of engineers and mechanics for whom they are not mere "devices," for whom, in fact, they are very beautiful. But this small group is not the source of the large-scale opposition to wind turbines.

28. Robert Thayer, *Gray World, Green Heart: Technology, Nature, and the Sustainable Landscape* (New York: John Wiley and Sons, 1994): 274.

29. It is clear from the prescriptions in his paper that Paul Gipe wants them to stay that way. Wind turbines should not expose themselves.

30. I believe, although I certainly cannot prove, that the transience of wind turbines, the fact that they can be taken down and set up anywhere in very short order, is a factor in the

resistance to wind turbines on the landscape. Again, if they survive in particular places over the long run, then their transient aspect will have been undermined and new possibilities for their appreciation opened up. Ours is a throwaway society, and it is part of the device-like character of contemporary technologies that they are disposable.

31. Christoph Schwahn catches just the right note: "Elements of technical civilization are very often standardized in their outfit. The more of them are placed into landscape, the less is the landmark effect. Because of standardization, wind generators can be very annoying in the marshes: formerly people could distinguish every church tower telling the name of the place. Today, wherever you look you always see the turning triblades. The inflation of standardized elements like high tension masts and wind generators puts down orientation and contributes to the landscape standardization caused by industrial agriculture."

32. Frode Birk Nielsen's video on Danish wind farms shown at the Villa Serbelloni workshop made this clear; the reaction of wind farm visitors was purely passive. In this respect, remote and opaque, they are like nuclear reactors, devices, although I would add that something is not simply a device or a thing. There are degrees.

33. Of course there are many exceptions. A Billings, Montana, doctor and good friend of mine, who has long fixed cars and airplanes in his spare time, decided after 14 years of frustration to learn how to maintain and repair his own three turbines. Fortunately, he now fixes ours as well. In fact, we have the only four regularly operating commercial wind turbines in the state of Montana.

34. Only very rarely do those who own the land have any sort of equity interest in the turbines. That it is easier to work with fewer rather than more landowners is another factor in the grouping of turbines. I very much applaud what is being done in Denmark and Germany to give local farmers an equity interest in and some measure of control over the turbines placed on their land. But I would add that to the extent that standardized machines are plunked down in a standardized way, then no matter who owns them, the *local* character of the community is thereby weakened if not also destroyed, and with it the possibility of feeling at home in it. To feel oneself at home in the world we first have to orient ourselves with respect to it, and this involves being able to distinguish between *things*.

35. E. F. Schumacher, *Small Is Beautiful: Economics as If People Mattered* (New York: Harper & Row, 1973): 33–34.

36. In a famous little essay, "On a Certain Blindness in Human Beings," in *Selected Papers on Philosophy* (New York: E. P. Dutton, 1917), William James notes the very different responses of a traveler, himself, and a local landowner to a forest clearing in the mountains of North Carolina. For the traveler, everything was visual, *scenery*, a "mere ugly picture on the retina," whereas for the landowner the clearing was "a symbol redolent with moral memories and sang a very paean of duty, struggle, and success." The point I (although not James) want to make in this connection is that we move beyond the visual (abstract and general) and merely scenic only when we make connection with local (concrete and particular) life, in which case the moral and the beautiful start to cohere. In their present anonymity, how can wind turbines make anything other than a visual impression (if not also an "ugly picture on the retina")?

37. Recognizing this has required knowledge of the way in which ecological units work and has led to well-organized attempts to defend the integrity of particular ecosystems and landscapes.

38. And are increasingly important in the determination of public policies.

39. Frode Birk Nielsen makes wonderfully clear that Danish turbines, in their native country, are "based on centuries of experience and tradition."

40. *Small is Beautiful*, 31.

41. We have experimented with various rotor diameters, from 20 to 70 feet (6 to 20 meters), all of them smaller than the towers on which conventional turbines are mounted. We have, in fact, gone back to a 20-foot rotor, which is small enough and simple enough that almost anyone can install it, unaided, on her own property.

42. "The windmill is another noteworthy feature of rural Portugal. Many windmills built centuries ago remain in use today. The most common is the picturesque Mediterranean type. The tapered cylinder of the tower is usually constructed of durable mortared stones covered with a finish of stucco. Always painted white, the tower is capped by a conical roof from which the mast protrudes. Usually the mast holds four triangular sails. When spinning with the wind, doing the work for which they were intended, the mills are a winsome sight indeed. Some farmers attach small clay jugs to the sail ropes. The small jugs whistle in the wind as the mill performs its task." T. J. Kubiak, *Hippocrene Companion Guide to Portugal* (New York: Hippocrene Books, 1989): 153.

43. No one, least of all those who directed California's largest utilities, foresaw the enormous volatility that deregulation would bring. It remains true, however, that deregulation opens the door to a great deal more small power production. Moreover, when the energy sources are renewable, as in the case of wind, then the sort of shock accompanied by the recent dramatic rise in the price of natural gas is dampened.

44. Three facts to keep in mind: (1) The world's largest technological–industrial companies have failed utterly in their (hugely well-financed) attempts to develop an efficient and reliable wind turbine. (2) Historically, the larger the turbine, the shorter its working life. (3) Robert Righter mentions in *Wind Energy in America: A History* that 5 million water-pumping windmills were at one time spread across the American West. At 1 kW per machine, they represented 5000 megawatts of distributed power where the risk both of machine failure and of wind failure was spread so widely as to be practically nonexistent.

45. I am grateful for the very helpful discussions of these issues, over many years, with Albert Borgmann, David Healow, Henry Kyburg, Robert Righter, and John Winnie.

PART

III

WIND POWER IN NORTHERN EUROPE

4

WIND LANDSCAPES IN THE GERMAN MILIEU

MARTIN HOPPE-KILPPER AND URTA STEINHÄUSER

The growth of wind energy in Germany has far outpaced that in any other country, with 1700 MW added in 2000 alone, bringing total generating capacity to nearly 6000 MW. The density of wind turbines on the landscape of Germany's most northern state is almost twice that in nearby Denmark, the country that pioneered the modern wind power revival. Not all Germans agree that this is a commendable development, some of them considering wind power's intrusion on the landscape tantamount to a catastrophe. Asserting that public acceptance is a matter of central importance in the further expansion of wind energy in Germany, Martin Hoppe-Kilpper and Urta Steinhäuser use several case studies to consider the proper reaction to turbines within the context of Germany's aesthetic consciousness, political realities, and legal mandates.

The use of wind energy in Germany has made enormous progress since 1990 (Figure 4.1). Initially, wind development was spurred by federal and state financial incentives, such as the "250 MW" research program. This federal program pays a subsidy for every kilowatt-hour generated by enrolled wind turbines. In return, the turbines' owners agree to regularly report on the operation of their machines, some of which are connected directly to a central monitoring system in Kassel via modem. The program is unique in the world and has produced a wealth of data on the performance of modern wind turbines. Yet this and the other early incentive programs were only modestly successful in spurring new installations. However, a decisive event occurred in 1991 when the *Bundestag*, or federal parliament, enacted the electricity feed-in (or feed) law (*Stromeispeisungsgesezt*). This law established the rate of reimbursement for electricity generated by renewable sources of energy that were fed to the national network. For wind energy, the electricity feed

Copyright © 2002 by Academic Press.
All rights of reproduction in any form reserved.

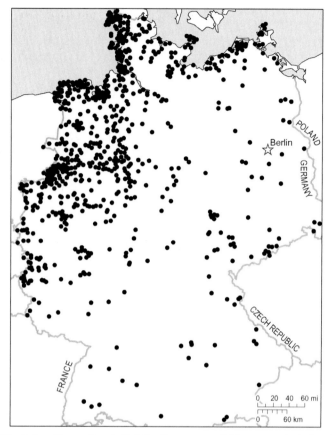

FIGURE 4.1   Generalized distribution map of German wind development.
Approximately 8000 wind turbines representing about 5000 MW of generating
capacity were in operation at the end of 2000. (Courtesy Institut für Solare
Energieversorgungstechnik e.V. [ISET], Kassel, Germany, http://www.iset.uni-
kassel.de/. Adapted by Barbara Trapido-Lurie, Department of Geography,
Arizona State University.)

law guaranteed that owners would receive 90 percent of the retail tariff for
electricity for every kilowatt-hour they generated. Within a few years,
Germany became the world's largest national market for wind turbines. As
a result of the feed law's unparalleled success encouraging new wind
development, wind turbines were producing about 2.5 percent of
Germany's electricity at the start of the present millennium.

   Although several studies of Germany's potential wind resources have
been inconclusive, they do confirm that there are sufficient resources for
wind energy to justify further expansion. According to these resource

assessments, wind energy could make up a significant percentage of Germany's electrical energy production. This would require a significant increase in the number of installed wind turbines, and it would inevitably stimulate new conflicts and land use debates.

Farmers and rural landowners pioneered the latest surge in wind development in Germany. From 1992 through 1997, about one-half of the operators in the federal government's 250 MW wind program were farmers. It is also likely that a large number of the shares in small wind companies are in the hands of farmers. There are several reasons for this. Farmers are in possession of prime properties. They also earn their living from nature and are not averse to landscape change. Farmers are also risk takers and willing to invest their own money in new crops or techniques. They have, in other words, a certain natural affinity for installing and using wind turbines on their land. At a time when European agricultural subsidies were being cut, German state and federal governments provided subsidy programs for wind energy development in rural areas in part to offer farmers an additional and welcome source of income. But the surge in wind installations has not come without its critics, including electric utilities and landscape protection societies. The electric utilities complain that under the electricity feed law they face an unfair burden of paying artificially high prices. Theirs is not a fundamental rejection of wind power, but rather a desire for a guaranteed method of compensation, preferably on a European-wide basis. If just compensation is agreed upon, criticism can be alleviated.

THE BASIC QUESTION

The reduction in the value of an existing landscape is the most frequent reason given for the rejection of a wind turbine building permit. It is an argument frequently used by those opposed to new wind energy projects. However, with careful thought, charges that new wind turbines reduce the value of a landscape can be shown to have questionable validity from the point of view of landscape management. With this in mind, the basic question is: How should we react to the growing criticism that wind turbines disfigure the landscape? Public acceptance will become a question of central importance in the further expansion of wind energy (Figure 4.2).

Often opponents to wind power attempt to make landscape an object in itself, an abstraction, without paying attention to the necessary work and conditions for its formation and its maintenance. It is reduced to a

FIGURE 4.2   Kaiser Wilhelm Koog. Schleswig-Holstein, Germany. (Courtesy Paul Gipe.)

backdrop for recreation. Recreation, however, is only one of many landscape uses we must address when dealing with public acceptance of wind energy.

Considered more completely, wind energy provides multiple benefits that accrue to society as a whole. It has positive effects on the energy supply, on industry, on agriculture, and on the environment. But these values are more global, whereas wind energy's impacts are local. Therefore, local concerns must be paramount in any project. Large-scale wind energy development is most successful when it is first *desired* at the local level, and only later valued by society as a whole. In several successful wind farms, including one I will later discuss, local understanding and acceptance preceded installation of the turbines. Such understanding includes the fact that the economic advantages accrue to those most directly affected, not only in a monetary sense but also in the ability to control a source of municipal power.

LANDSCAPE AND LANDSCAPE VALUES

In Germany, reactions to wind power have to be considered within the context of our landscape values. Generally, we believe that a landscape has a value of its own, something which must be protected from change. We treat it that way. Because of this concept, every new development requires special mitigation measures which often lead to compensatory levies paid to the local authorities for the perceived impacts. The federal nature conservation statute, Bundesnaturschutzgesetz, or BNatSchG, is the mechanism for these procedures.[1] The BNatSchG demands protection of both nature and the landscape to safeguard its variety, uniqueness, and beauty. This is a significant tool for landscape protection in Germany. Therefore, it is useful to analyze the origin of the terms and concepts introduced by the BNatSchG, and also how this statute is used in the current landscape debate.

Originally the term "landscape" was most closely linked to the visual arts, where the view was depicted two-dimensionally and bounded by a frame.[2] Put more concretely, a landscape view was considered a "representation of the landscape for its own sake."[3] Landscape artists of the Romantic period argued that their work expressed "the beauty of nature" or "the power of nature" with the help of selected landscape elements from the rural economy.[4] When landscape managers today make use of the terminology of the visual arts, they are obviously influenced by the Romantic's understanding and embellishment of the landscape. Artists of the Romantic period, then, have created our view, or concept, of landscape beauty.[5]

Beauty is not fixed, however, but susceptible to changing ideals. Even the ideal view of a beautiful landscape proves no exception. At the beginning of the 19th century it was the improvements in the rural economy, such as new techniques of cultivation or the construction of roads and pathways, which stood as symbols of order and human industriousness. These qualities were at the center of German philosophers' ideas about progress.[6] At the beginning of the 20th century, *wilderness* became the favored landscape ideal, synonymous with a natural setting largely devoid of human dwellings, and often areas with marginal economies.

The ideal wilderness was a concept completely devalued under National Socialism. Instead, values such as cleanliness, order, and dominion over nature gained prominence, and these values defined the view of the agricultural landscape until the 1960s. But with the rise of the environmental movement, the new ideal of an aesthetic landscape took shape. In the 1970s, interest in nature experienced a renaissance. In the present era, however, nature has been separated from the rural economy and the experiences which are linked to it. With ongoing changes in the ideals of beauty, landscape management steadily distanced itself from its own work as gardeners and landscape shapers.

Today, German landscape publications call the destruction of scenic beauty a "catastrophe."[7] The definition of landscape is reduced to only its artistic, tranquil, or contemplative aspects. Landscape becomes a visually experienced scene in a frame, just as in the landscape painting of the 19th century. In this case, landscape arbitrarily becomes synonymous with nature, where any connection between the outer appearance of the landscape and the economic conditions which produced it is negated.[8]

Yet in our industrial society, landscape is now usually defined as land developed and cultivated by humans. Contemporary landscapes result from anthropocentric influences; landscapes are created by people in the context of the prevailing rural economy. It could be said that without farmers who work the land sustainably, there would be no meadows and pastures, no arable fields, no enclosures, and no woodlands. In this context, lands that are developed and cultivated according to age-old practices become extremely valuable. These landscapes, however, are the products of a rural economy that is not economically competitive in a global marketplace, and, therefore, they are fast disappearing.

Discussions about landscape usually open with a statement of impending doom, such as: "The beauty of our landscape is in danger." The more this pronouncement is repeated, the greater the danger seems to become. The risk, then, must be counteracted by regulations, by a bureaucracy to turn away the threat. In the assessment that results from this process, the qualities defining the landscape must be outlined, and ways in which they can be aesthetically changed or managed must be addressed. Landscape specialists in the German government have taken up this task. The federal nature conservation statute sets the legal framework and legitimizes administrative action. The exact meaning of this law is seldom discussed. To make the aesthetic order manageable and understandable, landscapes have to be seen as *objects* in a scientific sense, even when we are dealing with a subjective value such as the visual perception of wind power. This squaring of the circle is carried out by several methods, including the so-called "landscape image analyses."[9] A scientific model imposed on a humanistic subject is seldom altogether successful.

Through such analysis, landscapes are taken from the subjective to the quantitative. Evaluators first divide perceptions into different elements, subjectively assess them, and finally transform them into a mathematical value. Beauty thereby becomes quantifiable, and the corresponding fees to compensate for the loss of landscape values can be calculated. Ironically, unexpressed and unrepresented in this process are urban dwellers in search of recreation, the primary "consumers" of scenic beauty. As urbanites flee from inhospitable living conditions, it becomes necessary to find rural escapes. Instead of taking on the difficult task of improving urban life, it is much simpler for planners to rely on rural landscapes as retreats. Such a policy maintains the status quo in the city, yet requires more from the rural landscape. Instead of revealing the underlying causes of the problems faced by urban dwellers, planners have elevated the urban dweller's urge for rural recreation to a basic human need.

On this theme, Werner Nohl writes that generally, "Landscapes are experienced as beautiful when their character meets the existential needs of the observers. Often, such landscapes have aesthetic effects on the observer which can be connected to his own hopes for a pure environment, homeland, peace, and liberty. Of course, such landscapes are not already a better world. How could they be? But they often help the observer to look symbolically beyond the limitations of the present and see the world to be better than it is."[10] Thus, for many planners, the landscape is primarily a holiday park for city dwellers seeking revitalization.

From the viewpoint of these planners, a landscape filled with wind turbines is a poor fit with the imagined need of urbanites seeking recreation. Therefore, special mitigation measures or compensation are required with wind turbine installation. Government administrators, and to a certain extent the nature protection associations, initially see every wind turbine as causing a negative impact on the landscape by reducing its aesthetic value. Every change is judged a deterioration. By German law any degradation requires special compensatory measures. To determine the manner and size of the compensation, the aesthetic value of the landscape must be studied, and the manner and size of the impact, particularly the reduction in value, must be calculated. The criteria developed to do this are as extensive as they are contradictory. Again, the decision comes down to a question of taste.

Any investigation about the relevance of landscape aesthetics must solve the question of positioning, or placement.[11] There is basic agreement that the best site for a wind turbine is either where the quality of the landscape has already been diminished, such as near buildings, or where the turbine can be hidden, such as at the edge of a woodland. In principle, however, the main environmental argument of the German government is that wind turbines are necessary, even though they cannot be reconciled with an ideal landscape.[12] As such, the government takes a positive stance, at the same time freely admitting that the development of wind energy will diminish the ideal landscape.[13] Not all organizations agree with this view.

THE POSITION OF THE BLS

The German Association for Landscape Protection, Bundesverband Landschaftschutz or BLS, is often at odds with the development of wind farms. Despite the fact that the name of the association suggests it is a main-line environmental group, BLS has devoted itself solely to criticizing the use of wind energy and blocking its expansion. This parochial association opposes the use of wind energy at specific sites by often-questionable methods.[14] The notable effectiveness of the BLS can partly be attributed to its successful lobbying of official landscape managers.

BLS demands compensatory measures under Germany's nature protection statute wherever wind turbines are proposed. They also insist that government planners take more direct responsibility for landscape management. As a rule, BLS succeeds in getting planners to order decorative measures such as the planting of copses, hedges, and extensive orchard meadows to mitigate the intrusion of wind turbines into the landscape. These look good and often help to give the area a more natural appearance. Yet, on the down side, such "compensation" is wasteful. Agriculturally areas are often taken out of cultivation and years of careful husbandry are lost. Moreover, these areas require constant care, and their maintenance is labor-intensive and costly.

Today, landscape planners must differentiate between what is beautiful and what is not. Although making this judgement is certainly not an exact science, they invoke their status as experts to make themselves appear indispensable. To accomplish their task, they have divided the countryside into zones: one zone requiring protection and another zone where certain uses are permitted. For purposes of analysis, valuation, and mitigation, then, the landscape has been divided up into "beautiful" and "ugly." Each parcel is considered separately, with the implication that the analysis has been exhaustive. In reality, however, the sum of the parts can never again equal a whole, for the whole is always more than just the sum of the parts. Human lives are affected by their interconnection and interdependence with the landscape. If these lives are ignored, then the whole is also destroyed. This happens when landscape is exclusively examined from an aesthetic or biocentric view, ignoring the anthropocentric or human occupation of the land. This returns us to a basic premise: German planners must not be unduly swayed by urban views of the landscape. They must consider the needs and traditions of rural residents as well.

THE EXAMPLE OF LANDSCAPES WITH EXPRESSWAYS

The preceding discussion stressed that landscape is as much an artificial construct, created by preservationists and managers, as it is a problem for the people who live near wind sites. We are assuming that when people assign value to a project they base their judgements mostly on the degree to which their own living space (economic, visual, acoustic, and hydrologic) is altered. To illustrate this assumption, we will compare public attitudes toward two highway projects which affected the landscape. We wish to compare a completed expressway, the building of the A44 expressway from Kassel to Dortmund in the 1950s and 1960s, with a currently planned project, an expressway from Kassel to Eisenach. The comparison is based on reports by the local Kassel press.

The Kassel–Dortmund expressway officially opened in the summer of 1975. The highway was first proposed in 1953. Political bodies and institutions of the Kassel region worked hard to include the expressway in Germany's highway program. These organizations expected economic benefits from new industries, new jobs, and increased tourism. Planning was completed by 1963, and construction began in 1966. In 22 years of reporting about the Kassel–Dortmund expressway, the question of the highway's negative impact on the landscape was never once mentioned. Various articles discussed how the highway was in harmony with the themes of a "beautiful landscape" and with "relaxation." Reporting on the opening of one stretch of highway, the Kassel daily newspaper HNA noted: "How charming this new expressway is, nestled in the beauty of the landscape of north Hesse. It was well thought out by planners." Upon the final opening of the completed expressway in 1975, the first signs of new environmental concern can be found in the lukewarm reporting that "the interests of the landscape and environmental protection had been fairly considered."[15] It is obvious that well into the 1970s the public saw there was no contradiction between "beautiful landscapes" and highways.

Clearly, the perception of beauty and of beautiful scenery is influenced by society's social and economic visions.[16] The Kassel–Dortmund express-way was a product of the German economic boom. It was linked to the still widespread idea of motorized travel as a symbol for well-being, comfort, progress, and a higher quality of life. With such positive expectations, the expressway could fit into a beautiful landscape.

How different this was from the controversy surrounding the express-way from Kassel to Eisenach, which has been hotly debated since 1989! In the intervening years a transportation policy dependent on a car culture has been criticized, and there is now general agreement that the individual automobile is a fundamental source of waste, causing pollution of air, water, and soil. Nearly everybody is directly affected by the negative impacts of motor travel, and even those who are economically dependent on motoring are aware of the direct environmental consequences of continued use of motor vehicles. So it is no wonder that the A44 expressway from Kassel to Eisenach is much more controversial than was the A44 from Kassel to Dortmund. What is most evident is the reaction of all the land owners in the highway corridor: none wants the highway routed past their own front door or "backyard."

Beautiful scenery and expressways, like oil and water, do not mix. Certainly, as earlier, there are expectations of benefits linked to highway construction, such as strengthening of the economy, more efficient distribution of goods, and affirmation of the general belief in progress. However, highways today also represent noise pollution, air pollution, damage to flora and fauna, and the depletion of soil and water resources. Most conspicuous, however, is that the appearance of the landscape is still low in order of importance in the Kassel–Eisenach expressway debates. In fact, in more than nine years of reporting it has never been mentioned even once.[17] This absence does not reflect a lack of concern for the landscape, but rather that the impacts on people, animals, plants, and the natural world are so direct and obvious that opponents of the project do not have to raise the question of the appearance of the landscape (Figure 4.3). Perhaps it is evident that whether an object in the landscape is linked to a sense of beauty, or at least a sense of goodwill, is mainly influenced by the connotations this object has in our minds and the expectations linked to it. In other words, when one is convinced or even enthusiastic about some-thing, one does not merely tolerate it, but can find it beautiful. Whether or not an object in the landscape stimulates a debate about aesthetics depends upon whether there exists a direct physical threat to people and the environment. If this expectation does exist, then the argument about the appearance of the landscape need not be introduced.

FIGURE 4.3   A planned expressway from Kassel to Eisenach. (Copyright by Jörg Lantelmé. Used with permission.)

IMPROVING PUBLIC ACCEPTANCE OF WIND ENERGY INSTALLATIONS

From the foregoing experiences, we can recommend three approaches for improving public acceptance of wind turbines. First, address political objectives and goals. Since the electricity feed law went into effect, the use of wind energy in Germany has grown by leaps and bounds. The reason is not hard to see and is not only due to the technological improvements and the increasing cost effectiveness of wind turbines. Legislation and incentive programs initiated by both federal and state governments make it clear that there is a political will to develop wind energy. Without exception, this has had a positive effect on public acceptance. The political objective of increasing the capacity of the two northern states of Schleswig-Holstein and Lower Saxony (Niedersächsen) has been especially beneficial. Each state wishes to install 1500 MW of new wind capacity by the year 2005. Planning regulations in these states reflect this objective.

The second approach is one that emphasizes continued technological development of wind equipment within the context of an active educational program. Manufacturers and installers have to minimize, as far as possible, the disturbance to people and the environment caused by wind turbines, including further reductions in noise emissions, improvements in component recycling, and the development of special nonreflective paints. When projects are first proposed, planners and developers must deal openly with the type and extent of possible impacts. Noise emission certificates, noise protection reports, shadow-flicker analysis, computer-generated visualizations, and ornithological studies must all be considered. A thorough evaluation of all possible consequences, along with an active educational program on the environmental benefits, can build public trust.

The third approach addresses the most common argument used by authorities in rejecting projects, namely despoliation of the landscape. Offering to pay token compensation for the "damage to the scenery" is, at best, a poor solution (e.g., 100 DM or about US$50 for each meter of tower height). The payment can have lasting effect by sending a negative signal to the public. Trade-offs of this sort suggest that rural peoples, those who create the landscape through their work and daily lives, are incapable of managing their own affairs. To get involved in the landscape discussion initiated by the German state authorizing officials involving statistics and compensation is to drive down a dead-end street. Instead, it makes more sense to talk about real impact, such as noise or shadow flicker, and to deal directly with the affected people regarding how much alteration of their immediate surroundings is acceptable. This discussion, however, must always pertain to specific locations and not be abstract.

CITIZEN PARTICIPATION

As an example of some of the ideas we have presented, let us look at the wind farm in Udenhausen-Mariendorf. This cluster of five turbines has been operating successfully since the mid-1990s. Local residents were involved in choosing and planning the location of the turbines, and participated directly by buying ownership shares in the units. Workers installed the five 600-kW turbines in the spring of 1996. The wind farm is located in the townships of Udenhausen and Mariendorf (Figure 4.4). The wind farm is incorporated as a company (GmbH & Co. KG) with limited financial liability and limited partnerships. The project began, as is so often the case, with the interest and activity of individuals. The current manager of the company has always been enthusiastic about wind energy, but several of his previous attempts at building a wind farm were unsuccessful. His earlier projects had been thwarted during the planning stage by opposition from local officials. When he met with a politician from Immenhausen who shares his enthusiasm for renewable energy, the conversation turned toward broader citizen participation in the project. Motivated in such a way, a group formed in 1994 and put up a small wind power plant with citizen participation. This group (the current stock-holders of the company) included a talented mix of experts on tax law, energy, environmental technology, and engineering. They all shared a

FIGURE 4.4   Udenhausen-Mariendorf seen from the south. (Courtesy Hans Georg Thiel. Copyright by M. Durstewitz. Used with permission.)

financial stake in the project, but they were also bound together by their interest in renewable sources of energy. So, the first step was proposing a wind energy project with citizen participation. These citizen shareholders then chose a suitable location, consulted with the landowners of the site, and assessed the attitudes of the affected local town councils.[18]

Once it was clear that the project would not fail because of administrative, political, or ownership problems, the group announced its plans to the public. Letters explaining the motivation of the proponents, details on the project—its technical arrangement (number, size, and performance of the wind turbines) and cooperative form of ownership—were distributed to all households in neighboring villages. Information about the possible disturbance to nearby residents caused by noise or shadow flicker from the turbines was made public from the start.

Although an informative trip to a town which already had an operating wind farm was poorly attended, participation in an investors' meeting in the town halls of Udenhausen and Mariendorf was much more successful, averaging 40 to 50 interested citizens. Although skeptics were few in number (the participants were overwhelmingly potential investors), the meetings offered a good chance to present the goals, intentions, benefits, and impacts of the proposed project.

INVESTMENT AND FINANCING

Crucial to the success of the project was the financial participation of a number of local residents. The total cost of the wind plant was 6.15 million DM (US$3.7 million). The shareholders invested 1.85 million DM (US$1.1 million) and the state of Hesse issued a grant for 1.47 million DM (US$0.9 million). The remaining 2.84 million DM (US$1.7 million) was financed with a loan from a German fund with revolving low-interest loans for environmentally beneficial projects. Of the 65 members in the cooperative, 23 were from Mariendorf, 7 from Udenhausen, 20 from Kassel (the capital of Hesse), and 15 others from the region of north Hesse and eastern Westphalia. A local bank arranged the financing.

The Udenhausen-Mariendorf experience demonstrated that raising sufficient capital only from small investors who purchase 2500 DM to 5000 DM (US$1500–3000) shares is difficult, if not impossible. Indeed, the participation of some large investors or the use of loans is indispensable. However, although shares of 2500 DM contribute little in an economic sense, they are successful in anchoring the project in the community. Obviously, a successful investment plan involves both outside capital and local investors.

Construction began in the early spring of 1996, with the plant coming online in April. After the erection of the turbines, locals could see for the first time how far a tower with a hub height of 53 meters (175 feet) reaches into the sky. It took some time for residents as well as participants in the project to become accustomed to the sight of the tall turbines on the hillside, a process of trust-building that succeeds most easily when the objectives of the participants and the benefits of the project are clear from the outset.

OPENING WITH "BIER, WIND, UND WÜRSTCHEN"

The wind farm was officially dedicated in May 1996 with a party at the site.[19] Publicity and the participation of the local residents remained an integral part of a successful program, not just as a means of completing the project. About 500 visitors came to the opening celebration to express their interest in the project and to enjoy the *Kaffee und Kuchen*, beer, sausage, and music (Figure 4.5). Since then, shareholders hold a regular "open house" at the wind farm every summer, giving both critics and supporters alike the chance to get firsthand information on the turbines, allowing them to make up their own minds about the project. Paul Gipe, another contributor to this book, has been advocating this welcoming approach for U.S. sites. At this site in central Germany, the interest has been lively, with between 20 and 30 visitors on such days, many climbing one of the wind turbines, a feat that is both physically and figuratively the high point of their visit.

FIGURE 4.5   Beer, wurst, windmills. Citizens gathered for the opening celebration of the wind project at Udenhausen-Mariendorf, Hesse, Germany. (Courtesy Paul Gipe.)

We would be less than candid if we did not point out that not everyone supported the project. Indeed, the mood in the villages of Udenhausen and Mariendorf can sometimes be hostile. There are also those in the villages who are not convinced of the need for the turbines. Still others have envious fantasies that the local owners of the turbines are making their fortunes at the expense of electricity consumers. But, as there is no noise disturbance at nearby houses and as shadow flicker has proved insignificant, opinions about the wind plant seem to be primarily based on attitudes toward energy policy. According to the project's shareholders, young people are more accepting of the wind turbines, more positive, and more interested than older people.[20]

SUMMARY

All in all, we know that wind plants provoke local debate. Much of this debate is healthy. How is our electricity really generated, and how should it be generated in the future? How much electricity do we really need, and why? How do I fit into all this, and how do I *want* to fit in? All these important questions should be discussed, and wind turbines invite commentary and participation in a necessarily democratic discussion about energy policy. This is as it should be.

Above all, the wind plant is a local enterprise. Local enterprises are never owned or operated by all of the people living in the vicinity, but rather by only a few. No project will ever win the support of everyone. Nevertheless, we have learned that a successful wind plant requires more than good wind resources and a good wind turbine design. A successful project requires:

- A political framework with government-supported programs and inclusion in building statutes
- Local decision-making (municipal-planning sovereignty)
- Interested and involved people in the project locale.

NOTES AND REFERENCES

1. Bundesnaturschutzgesetz (BNatSchG) Paragraph 1, Absatz 4, in der Fassung der Bekanntmachung vom 12.03.1987 (BGBl.IS.889).
2. *Microsoft Encarta 97 Enzyklopädie* "Bild" on CD-ROM, 1993–1996 Microsoft Corporation.
3. *Das große Lexikon der Malerei* (Braunschweig, Germany: Georg Westermann Verlag, 1982): 767.
4. See Caspar David Friedrich's "Einsamer Baum," 1822 (Berlin Nationalgalerie).
5. Jürgen Stolzenburg and Christine-Anna Vetter, Beitrag zur Disziplingeschichte der Freiraumplanung 1960–1980, in *Notizbuch 6 der Kasseler Schule*, 1983, Kassel, Germany. See also Andrea Appel *et al.* Ob Öko-, Deko-, Psycho-.... Hauptsache Grün: Ein Überblick über 40 Jahre Berufsgeschichte der Landespflege anhand von Fachzeitschriften, 1990, Kassel, Germany.
6. Ilke Marschall, Leitbilder fallen nicht vom Himmel: Zur Entwicklung und Bedeutung von ästhetischen Leitbildern des Naturschutzes und der Landschaftsplanung in Bezug auf die bäuerliche Kulturlandschaft, in *AG Ländliche Entwicklung, Fachbereich Stadt- und Landschaftsplanung der Universität Gesamthochschule Kassel: Arbeitsergebnisse*, Heft 35, 1996, p. 14, Kassel, Germany.
7. Werner Nohl, Beeinträchtigungen des Landschaftsbildes durch mastartige Eingriffe, Materialien für die naturschutzfachliche Bewertung und Kompensationsermittlung, p. 7, München, 1992. See also Sabine Schwirzer, Landschaftsverträgliche Windparks, in *Garten & Landschaft*, 8/1994, pp. 31–33, München, 1994.
8. cf. Nohl, Loidl. Schwirzer.
9. Hans J. Loidl, Landschaftsbildanalyse-Ästhetik in der Landschaftsplanung? in *Landschaft und Stadt*, 13 Jahrgang, 1981, Heft 1 (Stuttgart: Eugen Ulmer Verlag, 1981): 7–19.
10. cf. Loidl, Nohl, Gareis-Gramann.
11. Holger Brux, Windenergieanlagen und Landschaftsbild: Konfliktanalyse und Lösungswege, Beispiele aus der Praxis, in proceedings of Deutsche Windenergiekonferenz 1992, pp. 87–89, Wilhelmshaven, 1992.
12. BUND Landesverband Hessen e.V. Windenergienutzung in Hessen, Stellungnahme des BUND Hessen, 1995, Frankfurt.

13. Franz Alt, Jürgen Claus, and Herman Scheer, eds., *Windiger Protest: Konflikte um das Zukunftspotential der Windkraft* (Bochum, Germany: Ponte Press, 1998), p. 181 on statements of various environmental organizations to the use of wind power.

14. Alt, *Windiger Protest*, pp. 37 and 121. The Bundesverband Landschaftschutz (the German Association for Landscape Protection), or BLS, is an anti-wind group much like Country Guardians in Great Britain. BLS should not be confused with BUND, a much larger environmental group.

15. Hessiche Allgemeine und Hessich-Niedersächische Allgemeine, Archivierte Artikel zum Straßen-bzw Autobahnbau, aus den Erscheinungsjahren 1953 bis 1997, Kassel.

16. The first "autobahn swimming pool" in Germany was opened in Kirchheim, near the A7 in 1962. A vacation village was planned to follow.

17. It is to be expected that the theme would be dealt with in the expert reports, but obviously was of so little importance that it did not merit mention in the press.

18. Approximately 4000 square meters (about 40,000 square feet) of land area were bought for each wind turbine. Most of this land was needed to provide the required separation distance between the wind turbines and neighboring property. Some was also needed for access roads and foundations.

19. The enthusiastic description by Paul Gipe of the inaugural celebration of WEP Udenhausen-Mariendorf in May 1996 was posted on the Internet.

20. This observation is supported by a survey in the northern Black Forest. *Die Tageszeitung* (TAZ), No. 5476, 25.02.1998, p. 9, Berlin, 1998.

5

SOCIETY AND WIND POWER

IN SWEDEN

KARIN HAMMARLUND

Although sharp public responses to wind turbines are common, effective measures to set worries aside remain a matter of debate. Applying the results from her surveys of public opinion, Swedish geographer Karin Hammarlund argues that public opposition need not be the deciding factor influencing the future contribution that wind power makes. She believes that the key is careful public presentation of wind proposals and a direct appeal for early public involvement. No planning is really worthwhile without public participation.

WIND POWER'S PREDICAMENT

Each society is united by social institutions, institutions commonly slow to develop and slow to change. This presents a special predicament to wind developers, because wind turbines can alter the landscape more completely and more abruptly than any other type of land use. Less than a day is needed to erect a turbine, and the effects are visually immediate. This reality calls for a new dimension of planning. With visual changes to the landscape being not only quick but unavoidable, involving and preparing the public is an important step wherever new wind developments are planned. Such preparation must include the planning authorities. However, in Sweden, as in many other countries, wind energy has not been specifically considered during debate over national environmental and planning legislation. As a result, planners often treat the visual effects of wind turbines as an environmentally hazardous development. This is a serious mistake, for clearly there is a difference between a visual change in a landscape and an environmental hazard (Figure 5.1).

Copyright © 2002 by Academic Press.
All rights of reproduction in any form reserved.

FIGURE 5.1   Cooperation: From Häckenäs by Lake Vättern outside the city of Vadstena. Here cultural tradition and historic values coexist with present-day land use interests. (Courtesy Anne-Lie Mårtensson. Used with permission.)

Legislation is always open to interpretations based upon practical experience. In Sweden, the majority of environmental regulations mandate that existing uses should suffer no serious disturbance from subsequent developments. If the dominant presumption is that wind power will have a serious impact on the landscape, there is little chance for a successful project. I think that the intrinsic problems of planning originate from differences between experts concerning the approach to landscapes. The system in Sweden is one of "functional sectorization" in which different parts of the landscape such as nature, culture, and society are evaluated independently and therefore out of context. One of the effects of such a system is that land use is allocated by competition. For this reason, Swedish landscapes are constructed from power relationships and not a rational, balanced evaluation.[1] This is a significant obstacle to commercial wind development because successful introduction of wind turbines often depends on coexisting with functions and uses of the landscape that are already in place and upon which local residents often depend. There appears to be little recognition that wind power is a valuable ally of the landscape, one that can safeguard the long-term freedom of action in the landscape, a temporary guest that can leave without a trace.

Complicating the public's view of wind power, the changes it makes to the landscape are quick and obvious, while the personal benefits are invisible and only slowly realized. In contrast, planners and the public ignore the more gradual, albeit much more extensive, changes caused by farming because there is an explicit need to cultivate the land to provide food and because the landscapes remain nonindustrial and green. We tend to react to conventional energy systems in much the same familiar way because we have been living with them for many generations. Unlike wind power, they are widely distributed, and because they are close to urban areas they tend to be positioned within an industrial zone. In order to help us acquire a more balanced perspective on the sources of our energy, we need a policy which does not hide the long-term impacts of conventional energy systems, and therefore explicitly suggests the need for renewable energy. Such a policy will enable us to present a clear message concerning the environmental effects of our present use of energy. We may well reconsider the possibilities for wind power "in our own backyard." Indeed, there may actually be a wide national agreement on the benefits of renewables which national opinion surveys can verify. However, it is usually not possible to apply these results in order to guarantee local support for wind turbines because the support must come from the population directly affected.

THE CONCEPT OF LANDSCAPE

One of the challenges in trying to balance wind power with nature is, as Douglas Porteous concludes, that there is no solid consensus on the most useful aesthetic landscape quality appraisal methods.[2] I found this to be true when I participated in an official Swedish investigation concerning wind power siting called *Vindkraft i harmoni* (wind power in harmony) (Figure 5.2). One of our conclusions was that it is difficult to define general criteria for the location of wind turbines, because each landscape is unique. Indeed, there was such a lack of a consensus on the word "landscape" that we left it out of the title of the report, begging the question, "In harmony with what?"[3]

Obviously we need to define, or at least standardize, the concept of landscape. Landscape has a medieval Germanic connotation of an area belonging to and shaped by people.[4] In the Dutch concept of *landschap* which emerged in 1600, landscape meant the background of a portrait or a view of farms and fields. The social context was implicit. In 18th-century Britain, landscape became an aesthetic concept that could not be appreciated without appropriate training.[5] As we can see, the concept of landscape has a double meaning, either as a smaller territory with internal coherence, or as merely the visual surface of things which makes it almost indistinguishable from the term "scenery."[6]

FIGURE 5.2   Harmony: A landscape can be more or less sensitive to change. These wind turbines in Skärhamn on the island of Tjörn on the west coast of Sweden do not stand in harmony with local reactions. However, they seem to stand in harmony with the landscape. (Courtesy Anne-Lie Mårtensson. Used with permission.)

The discipline of geography has long focused on the interacting phenomena of landscape ingredients, including physical features and the attitudes and relations of political power. We are so accustomed to viewing and moving within landscapes that we blend the natural and cultural processes into something cohesive and meaningful. The local landscape is a daily practical reality in one way or another, and this reality must be managed as more than merely the visual surface of things.

One vital step in addressing the problem is to mobilize all senses.[7] If landscape is a cohesive and a meaningful totality in our minds, why is our management of landscapes so piecemeal? Our understanding seems to consist of scientific theories about the fragments as well as our personal experience of a totality that we know well as long as we do not have to explain it. However difficult, we can distinguish three categories of cultural landscape by use of research and epistemology.[8] The first is the classic approach of human geography, which defines the cultural landscape as the landscape as modified by human activities. The problem with this definition is that everything belongs to the landscape and therefore a categorization must be done to carry out a scientific study. This categorization inevitably reflects the values of the researcher.

The second category defines landscape as the environment upon which a value is placed. This is mainly a view of the landscape held by people working with the preservation of natural and cultural values. The problem here is determining the basis for valuing the different parts of the landscape. In the third category landscape is defined in terms of elements with special value to a particular group of people in a given socio-economic context. In this view, landscape is seen as something subjective, meaning that research concentrates on how cultural and social values are relative to a particular place. The problem here is that the same landscape is perceived and valued differently by different cultures.

Given the great disparity regarding landscapes, what is the best practical way to connect our personal everyday activities with the application of scientific knowledge? The Swedish geographer Torsten Hägerstrand suggests we use time and space.[9] We all need a place to be: we need space for our activities over a certain period of time. We all need pockets of local order, and our interests are bound to meet in the budgeting of space over time. Hägerstrand believes we should focus on the individual actors and their relation to the landscape over time. I have found it an approach with particular applicability in the context of wind power.

REACTIONS TO WIND POWER LANDSCAPES

My research has found different reactions to wind power among rural and urban dwellers. Farmers look upon wind generating equipment as a contribution to their rural subsistence. Farmers and other permanent rural residents in agricultural areas are accustomed to seasonal landscape change, change that reflects the dynamics of a living countryside. An innovation such as wind turbines which can add to the dynamics of the rural landscape might seem reassuring. Temporary summer residents, however, would not agree. Escaping the intense pace of the city, they are looking for recreation and recuperation in the countryside. They turn to such landscapes for the stability they offer. For such people, new wind turbines might not be a soothing or welcome change in the landscape, although some merchants find that the equipment of wind power lures tourist families stopping on their journey through the countryside.

These different actors view wind turbines in accordance with their personal relation to a specific landscape, and the amount of time they spend in that particular place. Similar differences between occasional and permanent observers can be drawn from wind developments elsewhere, such as Palm Springs, California. Accordingly, we can improve the chances for a constructive dialogue about landscape development if we can clarify the reasons why some people view wind power as a practical solution to sustainable development while others see it as a threat to landscape preservation.

Time is an additional factor when it comes to recognizing the effects of different developments. We tend to react more vociferously to change in the landscape than we do to widespread, perhaps even hazardous, but less visible environmental effects of development. Hence, if we summarize some important factors concerning the concept of landscape and how we view change, we find that time and space are the common denominators. We tend to view change according to custom of use, the pace of change, and the visual evidence.

Landscape design involves a process by which architects and planners try to be useful by taking into consideration ever-changing technical, aesthetic, and functional requirements.[10] If we let place, actors, and time structure the cultural landscape, we find that the age or the individual fragments do not decide landscape importance. Rather, it is the human occupant and his place in the spatial structure, over time, that equally help make the landscape both useful and beautiful. A landscape should be valued on its own terms: that is, on the basis of the conditions and the people that shaped it. In this way we relate our efforts to a particular place.[11]

PUBLIC INVOLVEMENT

The ideas and ideologies that have filtered through the historical layers of landscapes give them meaning and create functional patterns in our everyday surroundings. If we fail to recognize and consider these patterns, conflicts between different land use interests easily occur. In the beginning of my work with wind power in 1988, I was called in as a social geographer to examine the cause of problems that had occurred with public acceptance. I found that the central problem was not the wind turbines, but rather the management and planning process, which usually excluded the public. Hence, these wind power projects presented little or no understanding of the social landscape. It was as if it could be taken for granted that everybody would understand that wind power could fit easily into the pattern of existing land use. A decade later the issue of public acceptance is of central importance. In Europe the visual impact of turbines is the prime agent of negative public reaction. However, I believe that this is only the surface of a deeper problem. As experts provide more and more refined methods of visual presentation of sites and layouts, they do not solve a basic reality. The landscape is *a social arena*. This fact receives little attention. Consequently, the alienation of the public continues.

My research shows that involving the public in a wind power project has very little to do with public hearings about ready-made plans, especially when a landscape has been evaluated by experts. Individuals appraise landscapes in different ways and there are several preferences to be considered. I have found that the opinion about a project is often expressed by an engaged elite. By elite I mean a small group privileged by means, influence, or power in the local society: a group I call opinion leaders. These individuals do not represent the general public, although they might represent the strongest land use interest in the area. To rely on this elite group, however, is a mistake. If a wind developer wants to get the job done, he must consult with and consider the opinions of the "social landscape": that is, all people who will be effected by change.

If a plan recognizes how different people make use of the landscape, different values automatically become apparent. Then the question remains, whose opinions should be heeded? There will always be some individual interests which will be set aside. If a wind power plan clearly reflects local values, it is evident that there must have been a dialogue between different users throughout the planning process. I believe that if we approach the local population more directly and respectfully, they will help us to develop the full potential of different sites as well as safeguard future space for new development. Public acceptance is our best guarantee for a successful wind power development on land or sea. Interestingly, a major challenge in the future will be to define the population that will be affected by offshore wind development.

The fact is that the public is more prone to support a project they have had a fair chance to influence. I would even go so far as to say that in most cases, it is not carefully and aesthetically sited wind turbines that cause the main problem, but rather the manner in which a project is presented to the public. I think that we can all recognize the need to involve people in the process of change in their own neighborhoods. In a wind power context this means that we must establish a dialogue with people concerning how they make use of their surroundings and what they feel is important to protect in the landscape.

It is not possible to take everything into consideration when professionally designing a wind power site. It is, however, necessary to consider people's feelings when we enter their backyards and learn about the social network behind the sterile map. If a project has the confidence of the public there will be more space for artistic freedom and new solutions. The challenge is to use this trust in order to bring new meaning into a landscape. We cannot in the long run explain and defend the choice of location and design by saying that we used experts to help us anticipate people's social and aesthetic preferences. Instead we must consult directly with the people most affected.

TIMING AND VISUALIZATIONS

Presenting a wind power plan requires a sense of timing. In some cases, depending on the size of the project, it might be worthwhile to allow a certain period of adjustment. If people express misgivings, a large wind farm can be developed sequentially, making adjustments easier. Such adjustments should highlight the flexibility and reversible qualities of wind power development. Just because a so-called wind farm can be erected quickly does not mean that it should.

Today a lot of wind power projects initially use computer-enhanced photographs as aids to visualization, as Frode Birk Nielsen discusses more fully elsewhere in this book. These visualizations can cause problems with acceptance because still pictures do not present the true visual impact of wind turbines on a landscape. After all, the windmills will be turning. Neither do they present their functional contribution. People often depict wind turbines not as a source of renewable energy but as a new element in the landscape that will diminish its scenic value. On the other hand, visualizations of turbines undeniably have some value in accelerating social adjustment by providing an *idea* of what planned developments will look like. Inevitably, however, these pictures never truly depict the experience of an active wind turbine, although they are a great aid.

I have found that the benefits of using visualizations are enhanced by the presenter's professional training and his previous experience with wind turbines. If people can understand the rationale behind certain designs or if they can recognize some benefits in relation to other wind power locations, visualizations can work well to create a positive dialogue. In this context it is important to understand that a "picture" can both suppress the benefits of wind turbines and camouflage some of the visual effects. Hence, visualizations must always be accompanied by detailed explanations. We do not experience wind turbines only by seeing them, but also through hearing and feeling their presence. As we move through a landscape, things fall into place, and as we approach a wind turbine, we directly experience the force of the wind that is doing the work of turning the blades. The use of "virtual reality" should be a help in this regard.

My involvement and testing of visualizations convinces me that most people fail to relate to the fundamental thought behind aesthetic solutions. In 1997 and 1998 I tested several visualizations made by six different landscape architects. I asked representatives of the general public living in the areas concerned to grade the visualizations as good, acceptable, or bad in relation to how they harmonized with the surrounding landscape features. All at least made the grade of "acceptable." This result has to do with the relationship between form and function. Design that does not have an understanding of the human activities on the land will not connect to the functional pattern of the landscape. It will neglect the important recreational patterns or important viewpoints. It will not connect to the travel pattern of people, which is the way most people on a daily basis experience the landscape.

Landscapes possess meaning for people, and this meaning connects with how we make use of a place. This function strongly affects our conception. So, what a particular place means to me depends on what I do in that landscape. For this reason, I believe that the function of each particular landscape must be specifically integrated with the aesthetics and design of a wind power site. Form that connects with function will mean something to the affected population, and not just to the designer, planner, or landscape architect.

SETTING STANDARDS

The public represents a vital source of information on matters of development: matters which are not always apparent in land use plans. If a wind power developer provides information and actively solicits opinions, people are more likely to become engaged and there will be a corresponding increased sense of cooperation. Certainly communal ownership of wind turbines will increase cooperation. Some developments may prove feasible *only* if cooperative ownership is offered to those most

FIGURE 5.3   Options: It is not everywhere that wind turbines make energy choices as clear to us as at this site in the town of Svalöv in Scania, south central Sweden. (Courtesy Anne-Lie Mårtensson. Used with permission.)

directly affected. It is tempting for developers to skip involving the public at an early stage, fearing that such involvement will slow the project's progress. However true this might be, a project that does not meet with public approval in the final permit process will probably not get done at all. The loss in trust and negative public relations may prove much more costly and time-consuming than a well-conducted planning process.

Development of a wind power site is out of the question if it has not been socially anchored in the local society (Figure 5.3). Hence, to launch a large plan is a time-consuming and delicate matter in which not only expert evaluation but public cooperation is required. In the long run, it is more sensible for developers to adhere to new standards for landscape planning, rather than to resist the fact that there will be competition between different types of land use.

Wind power advocates must also accept some realities. For one, they need to recognize themselves as exploiters of the landscape, with impacts that are clearly more noticeable than, for example, a coastal nuclear plant. They must acknowledge that even uranium strip-mining and toxic waste disposal may not stimulate the same level of debate as the visual effects of wind turbines. We tend to ignore impacts which do not immediately affect our own neighborhood.

A MODEL FOR AGREEMENT

All this leads us to a point where we can come to some general conclusions concerning design and aesthetics in a social perspective:

- Landscapes will vary in their sensitivity to change. Such sensitivity depends upon many things, such as the structure and the accessibility of the landscape, and socioeconomic conditions.
- It is not fruitful to generalize people's experience of a given landscape. Feelings and reactions toward landscapes are strongly affected by local natural conditions, cultural traditions, economic circumstances, and individualism.
- In certain landscapes there is a long tradition of coexistence between a variety of land use interests; this can facilitate wind power location.
- Wind power can contribute to and even restore values in a landscape, provided we understand those values. In order to gain such understanding, we must be familiar with the history of the place as well as the present-day conditions.
- It is easier to explain the function of wind turbines if their design and deployment are related to existing industrial areas and buildings.
- The experience of a landscape is strongly affected by public access and the possibilities they see of making use of the landscape. Hence, the social qualities must be integrated with its visual qualities in order for us to be able to design and plan in harmony with the whole landscape.

THE SIMPLE TRUTH

All these conclusions connect to a single and simple truth: no planning is really worthwhile without public participation.[12] A plan can only work successfully if people can agree upon the concepts that guide the development and if the proposed development does not threaten their future access. Unfortunately, aesthetic landscape appraisal and evaluation are too often made by professional planners and consultants, independent of public preferences. Interview-based preference methods used in sociological surveys can be quite helpful if we are looking for ways to get information from the public and to conduct a dialogue. In this dialogue we must sharpen our arguments concerning the benefits of wind power in order to answer the question of why turbines are to be located in a particular place.

We must also present ways for individuals to benefit from wind power if we expect their acceptance of such an intrusion on the landscape (Figure 5.4). No matter how obvious it seems that our reliance upon nonrenewable energy sources must eventually end, it is not clear to everybody that this situation will demand something out of us all. In the long run it will become more and more evident that the greater control that a society has over its supply of energy, the greater will be its total control over its own destiny. A decentralized energy system based on renewables will allow greater independence, less vulnerability, and more

FIGURE 5.4   Restoration: It is easy to manifest the function of wind turbines if they are related to existing industrial areas. In this case the wind turbines contribute to the restoration of an industrial area in Skärhamn, on the island of Tjörn on the west coast of Sweden. (Courtesy Anne-Lie Mårtensson. Used with permission.)

responsibility by bringing the sources of that energy closer to the individual.

We know that our response to turbines is formed quickly, but we tend to forget that the total benefits lie hidden in the future. It is important to widen the discussion concerning the effects of wind power to include all our senses in the planning process. I think it will be hard to resolve the aesthetic impact of wind power if we do not recognize that what we are dealing with is mostly an ideological discussion. How will the public respond to the question of whether they are prepared to accept an energy system based on extensive use of renewable energy sources? If the answer is "No," we must be certain that they have a clear understanding of the negative and irreversible effects to all life from the continued use of fossil fuels and nuclear power.

Are we aware of the full effects on our landscapes from our present energy systems? I think not. If we were, discussions concerning wind power would not tend to center on the visual impact of turbines. Aesthetics enrich our lives, yet we must make sure that we can stay alive to enjoy such pleasures. The question should not be whether to use wind energy, but rather how we can use it in the best way.

REFERENCES

1. Karin Hammarlund, "The social impacts of windpower," proceedings of the European Wind Energy Conference EWEC 97, Dublin, 1998, pp. 107–114.
2. J. Douglas Porteous, *Environmental Aesthetics: Ideas, Politics and Planning London* (New York: Routledge, 1996): 208–209.
3. *Statens Energimyndighet, Vindkraft I Harmoni* (Malmö, Sweden: Tryckeritecknik, 1998), in Swedish.
4. M. Jones, "The Elusive Reality of Landscapes: Concepts and Approaches in Landscape Research," in *Norsk Geografisk Tidsskrift*, 1991, Vol. 45, pp. 229–244.
5. Porteous, *Environmental Aesthetics*, 47–48.
6. K. R. Olwig, in *Sexual Cosmology in Landscape, Politics and Perspective*, B. Bender (ed) (Oxford: Berg, 1993): 307–343.
7. Porteous, *Environmental Aesthetics*, 41.
8. M. Jones, "The Elusive Reality of Landscape: Concepts and Approaches in Landscape Research," *Norsk Geografisk Tidsskrift*, 1991, Vol. 45, pp. 214–242.
9. T. Hägerstrand, *Time Geography: Focus on the Corporality of Man, Society, and Environment, in The Science and Praxis of Complexity* (Tokyo: United Nations University, 1985): 193–216.
10. Steen Estvad Petersen, "The Music of the Landscape," preface to *Wind Turbines & the Landscape: Architecture & Aesthetics* (Århus, Denmark: Birk Nielsens Tegnestue, 1996): 5.

11. T. Germundsson, and M. Riddersporre, Landscape, Process, and Preservation, in *Landscape Analysis in the Nordic Countries: Integrated Research in a Holistic Perspective*, The Swedish Council for Planning and Coordination of Research, Report 96:1, 1996, pp. 98–108.

12. Porteous, *Environmental Aesthetics*, 240.

6

A FORMULA FOR SUCCESS
IN DENMARK

FRODE BIRK NIELSEN

The care used in the development of wind power significantly affects not only how well the turbines are balanced with nature, but how the public reacts to the technology. Today, through sensitive integration of landscape values and the incorporation of computer visualizations, wind turbines have been installed in the Danish countryside and offshore with substantial public support. The Danish approach to wind energy development has helped Denmark move closer to its commitment to greater energy independence and responsible power generation. In following such precepts, Denmark is not only on a path to producing more of its own electricity, it is creating a model for wind development everywhere.

So slightly does Denmark rise out of the water that the wind's strength hardly diminishes between the North Sea and the Baltic Sea. With this resource so available in a country poor in other sources of power, Denmark has endeavored successfully to put its wind to work, ever careful to balance its demand for energy with the need to protect the natural and cultural attributes of the land. From the end of the 1970s to the early years of the next decade, wind turbines were usually erected in solitary installations. Gradually, the arrangement and pattern of wind turbines changed from individual, punctiform installations on the land to spatial installations with a directional and linear nature. This change in form, function, and scale has prompted new reactions in the countryside. Significantly, it has increased the number and variety of locations to evaluate wind power developments relative to landscape design.[1]

Landscape appearance and proportion always change with the erection of major structures. With the aim of evaluating how a structure is best

Copyright © 2002 by Academic Press.
All rights of reproduction in any form reserved.

adapted to a given landscape, one must weigh the pros and cons from an aesthetic point of view. This includes many contradicting and subjective factors, such as the production of clean electricity and the resulting symbolic value of wind turbines, and attitudes toward nature and landscape, as well as tradition. The goal is to establish a symbiotic relationship between the structures and the water or land surfaces: a visual balance or a unified whole created by the turbines and the natural elements of the landscape (Figures 6.1, 6.2a, 6.2b).

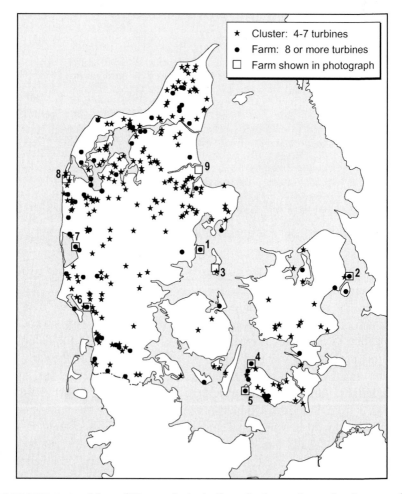

FIGURE 6.1   Map of Denmark, including all places shown in photographs. 1, Tunø Knob. 2, Middelgrunden. 3, Samsø Island. 4, Vindeby. 5, Kappel. 6, Tjæreborg. 7, Velling Mærsk. 8, Klinkby. 9, Overgaard Gods.

FIGURE 6.2   (a) Klinkby. The cluster of four turbines near Klinkby in Northwest Jutland (about 5 km west of Lemvig) is an example of a smaller installation with architectural conviction, via location and design, and in fine balance with the surrounding landscape. Workers erected four turbines on a gentle, raised plateau at the edge of the valley which underlines and connects the installation, forming a visual basis for it. The landscape is gently rolling with scattered farms and a number of bronze-age burial mounds, bordered by a winding stream channel to the west. Parallel to the row of turbines, a transmission line strung on wooden poles (H-frame) crosses the valley. The tight spacing between turbines (3.7 rotor diameters) adds to their appearance as a solid, cohesive composition with presence and authority. This is a unique and harmonious example of how small arrays of turbines can often be tightly packed because the interference of one turbine with the next is relatively low for small groups.

The landscape is the starting point. For the attentive observer, the landscape with its shapes and contours will suggest the direction and extent of development. Its character, structure, and topography should first be analyzed, and its signals used to form the basis of any proposed project.

A wind farm or group of wind turbines is like a gigantic sculptural element in the landscape, a land-art project. The actual design, spacing, height, type of wind turbine, and surface treatment of the sculpture must depend on the potential of the landscape in question. We must make the wind turbines and the landscape a coherent unit emphasizing both the

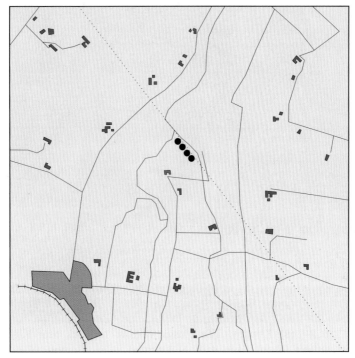

FIGURE 6.2   (b) Map of Klinkby showing placement of turbines from Figure
6.2 (a).

natural and the man-made. Apart from the aesthetic aspects, the following
functional aspects must also be considered:

- For optimal functioning, wind turbines must be erected so that they
  intercept the wind.
- Turbine spacing must be such that the turbines themselves do not
  greatly obstruct the flow of the wind from one to the next.[2]

The scale of wind turbines, especially in flat terrain, often exceeds all
other elements in the landscape. Moreover, in order to utilize the best wind
conditions and thus to optimize production, wind turbines are located in
exposed positions in the landscape. Here, form and function become
inseparable elements.[3] The only practicable way to achieve a result that is
positive both visually and functionally is to accept the fact that large wind
turbine installations are dominant units in the landscape, visible over great
distances. This, however, does not mean that the landscape must be
visually overwhelmed. On the contrary, a well-planned location for the
wind turbines can enhance landscape contours and contrasts.[4]

VISUAL ORDER

In the design process certain overall aesthetic considerations are worth remembering. For example, order is the first commandment of aesthetics. It is important that when locating an array of wind turbines, they should be seen as a clear coherent unit: that is, in geometric, often linear formations, in contrast to the landscape. At the same time, it is essential that a wind farm be delineated in a clear, unambiguous, and simple way, both at close range and from a distance. This is best achieved by giving the wind farm or wind power plant an identifiable shape, for example, as a closed system with a quadratic, rectangular, or triangular form, and by creating rhythm and order in the internal geometry. To properly express this form sufficient space is necessary. There must be a significant distance from the wind farm to other wind turbines in the area (Figures 6.3a, 6.3b).

Second, curved lines present particular design challenges because they can be difficult to distinguish at a distance. At the same time, however, the given formation of the landscape can underline and accentuate such forms, and thus can justify curved lines in special situations (Figures 6.4a, 6.4b).

A third consideration in the landscape architecture of wind power is that the appearance of a wind farm should be simple and logical, thus avoiding visual confusion, at the same time underlining the character of the man-made element.

Fourth, wind turbines located in flat and open terrain, such as exists in much of Denmark, underscore both the land and the wind turbines themselves. The vertical appearance of the wind turbine towers forms a contrast to the flat landscape, thus accentuating the horizontal aspect. Wind turbines located in a landscape already featuring vertical elements may result in a blurring effect. Where wind farms are located in flat and open landscapes, the retreating rows of wind turbines in a wind farm create perspectives that reveal the depth and distance of the landscape. When we erect wind turbines in geometrical order, such as in rows or modular networks, this open space perspective stands out even more clearly. Here it is essential that the individual turbines be located in accordance with an overall, thorough-going system so that we perceive the wind turbines as a coherent cluster rather than single, scattered units.

OFFSHORE WIND FARMS

For land installations there are practical limits to the size we can use.[5] Marine areas, however, provide a unique opportunity for a great number of

FIGURE 6.3   (a) A computer visualization of 25 units, 2.0-MW wind turbines on the estate of Overgaard Gods, Denmark. (Published in Frode Birk Nielsen: Vindmøllepark ost for Overgaard Gods. Birk Nielsens Tegnestue for Jysk Vindkraft, Aarhus, Denmark, 1998, p. 24.)

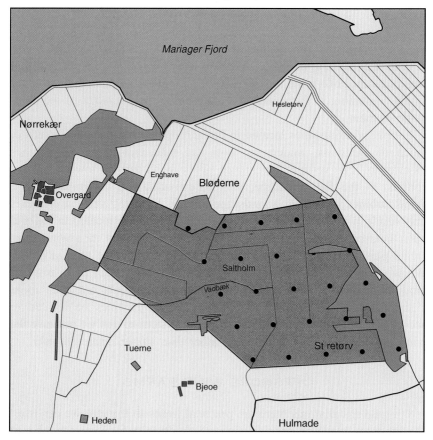

FIGURE 6.3   (b) Map of Overgaard Gods showing placement of turbines from Figure 6.3 (a).

FIGURE 6.4   (a) Kappel: 24 turbines (9.6-MW installation), commissioned August 1990. Southwest Lolland has ideal conditions for wind energy. By the late 1990s four wind farms had been erected on Lolland, an island between Germany and Denmark's largest island, Zealand (Sjælland), where Copenhagen is located. The area near Kappel is diked, flat, and open. The polder landscape features many drainage channels. The 24 wind turbines of the Kappel wind farm are erected in a single, compact row (turbine spacing of only 3 to 4.5 rotor diameters apart) that follows the coast and the gently curved course of the dike. The wind turbines are located directly behind the dike, connected with the gravel access road and anchored on a concrete pad. The wind turbines fit well into this intensively cultivated landscape and significantly emphasize the coastline, with the dike as a visually connecting element for the row of turbines. Local government granted an exemption to allow construction close to the beach. Irrespective of where you stand, the construction is visually strong and in harmony with the surroundings. The Kappel installation is a good example of how curved lines of turbines can be well suited to certain landscapes.

very large turbines.[6] The Danish Ministry of the Environment and Energy's committee on offshore wind turbines has recommended five areas in Danish waters which are sufficiently large for a major wind farm, and where there are no competing interests. In theory, these five areas could host approximately 3500 turbines of 2 MW each, with an expected annual electricity production of 15 to 18 TWh (15–18 billion kilowatt-hours), corresponding to about 50 percent of Danish electrical consumption.[7] The first of the five wind farms will be erected in the summer of 2002. The project will consist of eighty 2-MW turbines placed in the

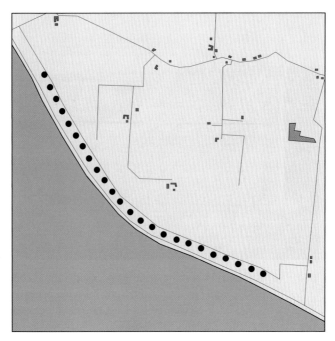

FIGURE 6.4   (b) Map of Kappel showing placement of turbines from Figure
6.4 (a).

North Sea, 40 km (25 miles) from Esbjerg, a port city on the west coast of
the Jutland peninsula. The turbines will generate enough electricity to
meet the needs of 150,000 typical Danish households (Figure 6.5).

Offshore placement of wind turbines has already begun. The world's
first offshore wind farm was erected in 1991 near the village of Vindeby
on the island of Lolland, where eleven 450-kW turbines are aligned in two
parallel rows (Figure 6.6). This project was followed by an installation of
ten 500-kW turbines at Tunø Knob in 1995, off the east coast of the
Jutland peninsula near Aarhus. Since then, two demonstration wind farms
have been installed in shallow waters in the Netherlands, and workers have
constructed several small pilot projects off the island of Gotland in
Sweden.

The visual consequences of offshore locations are different from those
that occur on land. Characterized by an unobstructed view, offshore
turbines can be seen over long distances, depending on visibility and
the play of sunlight on the turbines.[8] Based on experience from Denmark's
Vindeby offshore project, the power company has concluded that there are
no real problems—only advantages—in terms of environmental and
public acceptance of offshore siting.

FIGURE 6.5   Visualization of Horns Rev offshore wind farm. Seen from a distance at about 8 km. Project developer: ELSAM A/S. Landscape architects: Birk Nielsens Tegnestue.

As far as the wind turbines' impact on marine life is concerned, studies have shown that wind turbine foundations lead to better seabed conditions for the organisms that support fish and thus for fish stocks. A detailed bird study at Denmark's Tunø Knob offshore wind farm shows that most birds ignore the turbines and simply go where there is food. From this experience, offshore turbines seem to improve conditions for both fish and birds.[9]

New offshore projects are in the offing. During the late 1990s preliminary work was underway to install an offshore wind farm in the Øresund, between Copenhagen and Sweden. This project is cooperatively owned and independent of any government programs. By the end of 2000 the project was completed. The turbines are located just east of the Danish capital in shallow waters known as Middelgrunden. Under good conditions the turbines are visible from the parliament building, Christiansborg, as well as from the coast of the metropolitan area. The turbines are owned cooperatively by investors living in the city of Copenhagen. The organizers of the cooperative already operate the Lynetten wind farm within Copenhagen's harbor, visible from the parliament building and the Little Mermaid, a popular tourist attraction (see figure in introduction).[10]

FIGURE 6.6   Vindeby: 11 (4.95-MW installation) turbines. The world's first offshore wind farm was commissioned 2 kilometers off the north coast of Lolland in September 1991. The 11 wind turbines stand in shallow waters 3 to 5 meters (10–20 feet) deep and are oriented in two parallel rows trending in a northwest to southeast direction, transverse to the prevailing winds. This simple pattern is easy to perceive from all angles, and the perspective corridor created by the two rows of columns appears dramatic as you pass by. The wind turbines form part of an intimate visual interplay with the coastal landscape. In bright sun, the wind turbines are easy to see from the shore, appearing white. In overcast weather, however, they assume a grayish cast, which significantly reduces their visual dominance.

Another area of offshore activity may be in waters near the island of Samsø. The Danish Ministry of Energy selected Samsø in a nationwide competition to test the feasibility of using 100 percent renewable energy.[11] An important element in the plan is the establishment of an offshore wind farm with about ten 3.0-MW turbines. The wind turbines are intended to produce twice as much electricity as Samsø currently consumes. The surplus production from the offshore wind turbines will be used to displace fossil fuels now being consumed on the island, especially in transportation. Though liquid fuels can be produced from renewable sources, Samsø will use much of the surplus electricity to power electric vehicles. The project has won local support as an opportunity to inject new life into a static economy.

Part of the process of public approval was the use of visualization in the review process. In accord with recommendations that have been made elsewhere in this volume we believe a detailed visual evaluation is essential for properly siting wind turbines. Such a visual assessment of the aesthetic expression of the installation should, therefore, be completed prior to determining the exact placement of the wind turbines in the landscape.[12] Here, visualization is a means for projecting and assessing the consequences of the actual technical installation. Various methods and techniques are available which can be used at the planning stage to visualize a future wind farm or cluster of turbines: drawings, photo montages, computer-generated displays, and moving pictures from video or film.[13] Thorough visualization is an essential part of the democratic process and public outreach. It can make it possible to see exactly what the wind turbines will look like on the landscape, and how both the public and the neighbors will likely be affected.

During the early years of wind development, little attention was paid to the importance of visualizations. Later, after facing a critical public, European developers gave landscape architecture a higher priority. Today it is common practice on major projects to employ a landscape architect at an early stage. The architect draws up sketches with proposals for the number of turbines, their exact location, their relative positions, height, and so forth. This is followed by visualizations of the overall landscape composition, viewed from various distances and angles with the idea of providing a realistic picture of the whole complex. The aim is both to correct any undesirable or unharmonious effects at the planning stage, and to give each citizen a realistic picture of what the future wind farm will look like (Figures 6.7, 6.8, 6.9).

MODERN WIND TURBINE DESIGN

Wind turbines have resurfaced over and over again throughout history. As if proving that there is nothing new under the sun, the present design stage of Danish wind turbines recalls a forgotten past when, for centuries, wind was used as a source of power. Their design has continually improved, if inconsistently, based on past experience. It does not seem at all illogical that aircraft technology has been an essential source of inspiration in the design of modern wind turbines.

When looking at a typical modern Danish wind turbine, its appearance demonstrates sound aesthetic design principles: the tower is a round (or polygonal) metal structure, slim and conical. On top of this sits the moveable aerodynamic nacelle, with its hub, main shaft, generator, gear-box, and controls. Finally, the turbine has three fiberglass blades attached to the hub. The rotor, the combination of blades and hub together, is upwind of the tower, that is, it always faces the wind thanks to a computer-monitored wind vane. Tower, nacelle, and rotor are painted white or pale gray, and perhaps provided with a nonreflective finish. It is important that wind turbines in wind farms both offshore and on land appear uniform

FIGURE 6.7   Visualization of a suggested offshore wind farm near Samsø. The original idea was to place ten turbines in a circle array. The project is not yet realized, but is expected to be built in 2003 in a linear array (see Figure 6.9).

with respect to each other. Their overall configuration, color, and height should be similar. The rotors of the wind turbines should also have a uniform diameter, direction of rotation, and speed of rotation (Figure 6.10).

By the start of the new millennium more than 6000 electricity-generating wind turbines had been erected in Denmark over a period of nearly two decades. Some of these wind turbines have, of course, given rise to heated public debate and opposition. However, the majority have received positive support. This was due mainly to wind power's origins as a popular or grassroots movement. People supported alternative sources of energy as part of their determination to create a cleaner environment. Originally, the drive for alternatives was part of a widespread Danish resistance to nuclear power. Later support came from people who wish to phase out fossil fuels and thus reduce CO_2 emissions. Visualization was

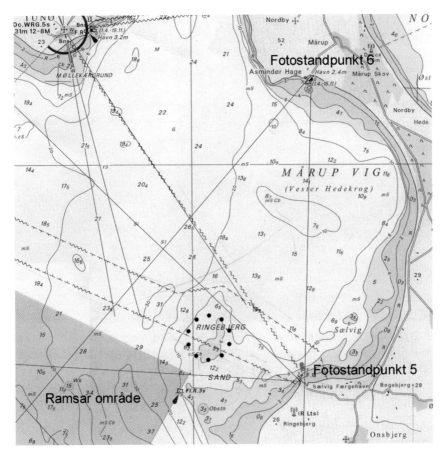

FIGURE 6.8   Map of suggested wind plant off the island of Samsø.

used throughout the process of public review of the wind power alternative.

PATTERNS OF OWNERSHIP

Many of the turbines in Denmark are individually owned, or owned by cooperative associations. Consequently, a large number of Danes are not only socially, but economically committed to the operation and dissemination of wind power. The public's involvement and investment in wind energy has been a crucial factor in its expansion. A wind turbine

FIGURE 6.9   Visualization of offshore wind farm near the coast of Samsø.
The project is proposed to be built in 2003. Project developer: Samsø Energi-
selskab. Landscape architects: Birk Nielsens Tegnestue.

cooperative (for example with three to five turbines) is typically made up
of several hundred small investors, all of whom can note with pride that
they have made a good investment and have a lower electricity bill as a
result. There is a residency requirement for participation in a wind turbine
cooperative. Thus, only citizens of the district where the turbines are
located or those in the adjacent districts can invest in the cooperative.
There is also a limit on the amount any one investor may own in a
cooperative wind turbine. These provisions guarantee decentralized
ownership. Big, absentee investors are kept out.

Only in the 1990s have Danish power companies played a significant
role in the expansion of wind power. The utility companies preferred to
build larger wind farms than the cooperatives: for example, 20 to 50
turbines instead of the small clusters typical of co-ops. However, several of
these utility-sponsored projects encountered strong local resistance and
were abandoned. So the power companies, encouraged by the government,
have now turned their gaze toward the sea, where fewer private interests
are involved, and where the wind resources are better than those on land.

FIGURE 6.10   NEG-Micon's 1.5-MW turbine. The general appearance of the turbine was designed by renowned Danish industrial designer Jacob Jensen and erected near Tjæreborg in 1995.

Wind power has grown substantially in Denmark in recent years. In 1979 I drew up a proposal for four wind farms, one of which was located offshore. Together they were projected to produce 10 percent of the country's electricity consumption at that time.[14] Although critics claimed that such a goal was totally unrealistic, Denmark has surpassed that target, and with one-tenth as many turbines as expected. The government's 1981 energy plan estimated that approximately 60,000 wind turbines would be required to meet 10 percent of Danish electricity consumption.[15] Today, as turbines have become both larger and more productive, we know that far fewer will be needed.[16]

In 2001 the 6500 wind turbines produced about 15 percent of the country's total electricity consumption.[17] Most of these turbines are 100- to 400-kW units, and were typically erected in the 1980s. More recent turbines are 600- to 900-kW capacity, and 2.0-MW turbines have been introduced. Through gradual replacement of the old by larger, present-day units, it is estimated that wind energy will be able to provide 30 to 35 percent of Denmark's electricity consumption. Further offshore wind turbines would add significantly to the total.[18] The government's official Energy Action Plan expects 50 percent of the country's electricity will be met with wind power by the year 2030, resulting in the highest use of wind energy of any industrial nation in the world.

The ebb and flow of domestic wind energy can be coupled with the existing hydroelectric power system in Sweden and Norway. This will enable Denmark to balance the availability of wind energy with that of hydroelectric power of those more mountainous countries. Excess wind energy in Denmark will offset hydroelectric generation elsewhere in Scandinavia. Effectively, the energy from good wind years will be stored as water in reservoirs behind dams. The water can then be released when wind turbines cannot meet their share of electricity consumption in Denmark. This exchange of renewably generated electricity will ensure that there is neither a glut nor a shortage of power in Denmark caused by fluctuations in the wind.

ACCEPTANCE OF THE DANISH WIND TURBINE LANDSCAPE

If, for some reason, we were to remove Denmark's 6000 wind turbines, there would be a public outcry. Wind turbines are now seen as an integral part of the Danish cultural landscape. They are viewed as a physical manifestation of our collective wish to reduce pollution. Our streams are no longer clean. Our forests are affected by acid rain. If I, as an individual, have a choice between the visual intrusion of a wind turbine and the physical pollution of a fossil-fuel plant, I would prefer—even as a landscape architect—the visual pollution. It is not, of course, my decision alone. My choice in a democratic society is that we Danes construct a power system principally based on wind energy, regardless of whether wind turbines are placed in single units, in pairs, in clusters, or in large wind farms. However, and most important, we must develop wind energy with variation, imagination, with originality, and in harmony with the surroundings.

For centuries wind turbines and windmills have been a characteristic and sometimes dominant element in the Danish landscape. They helped us protect our land and our landscapes from persistent and often pernicious pollution. In the future wind power should continue to be developed and expanded. Turbines, when located thoughtfully and sensitively, can enrich the cultural landscape and be an integral part of it.

NOTES AND REFERENCES

1. Frode Birk Nielsen, *Wind Turbines & the Landscape: Architecture & Aesthetics* (Aarhus, Denmark: Birk Nielsens Tegnestue, 1996). (Published with financial support from the Danish Energy Agency's Development Program for Renewable Energy.)

2. Spacing between wind turbines normally ranges from three to seven rotor diameters, depending on turbine size, the number of units, and their spatial pattern.

3. The energy in the wind is directly proportional to the cube of wind speed. The location of a wind turbine at an optimum wind speed site is, therefore, an extremely critical determinant of the amount of energy that can be produced. For example, if one site experiences twice the wind speed of another, the windier site contains not merely twice the energy but eight times more, that is, $2 \times 2 \times 2 = 8$.

4. Frode Birk Nielsen, "Landscape Aesthetic Considerations," *A Guide to Wind Farms I, Wind Physical Planning* (Vejledning i mølleparker I, Vindfysisk planlægning). (Hurup Thy, Denmark: 1984): Folke Center for Renewable Energy [Nordvestjysk Folkecenter for Vedvarende Energi], 150–159, in Danish.

5. Ministry of the Environment (Miljøministeriet, landsplanafdelingen), Spatial Planning Department, *Wind Turbines in Municipal Planning* (Vindmøller i kommuneplanlægningen), 1994, in Danish.

6. Wind conditions over areas of water are considerably better than over land for two reasons. First, the average wind speed over unobstructed water is typically 1 m/s higher than over land, resulting in approximately 40 percent more wind energy. Second, the wind is less turbulent over water than over land. This also contributes to increased wind turbine performance.

7. Danish Energy Agency (Energistyrelsen), *Wind Turbines in Danish Waters: A Survey of Authority Interests, Evaluations and Recommendations* (Vindmøller i danske farvande—kortlægning af myndighedsinteresser, vurderinger og anbefalinger), 1995, Copenhagen, in Danish.

8. Danish Energy Agency (Energistyrelsen), *Wind Turbines in Danish Waters: An Analysis of the Visual Impact for Erecting Off-shore Turbines* (Vindmøller i danske farvande—en undersøgelse af de visuelle forhold ved opstilling af vindmøller på havet), in Danish, 1994, Copenhagen.

9. I/S Midtkraft, *Tunø Knob Wind Farm: Visualisation and Aesthetic Evaluation* (Tunø Knob vindmøllepark—Visualisering og æstetisk vurdering), 1994, in Danish.

10. The wind turbine association of Middelgrunden and the electricity services of Copenhagen (Middelgrundens Vindmøllelaug & Københavns Belysningsvæsen), Wind Farm Located on Middelgrunden II: Aesthetic Evaluation and Visualisation (Vindmøllepark på Middlegrunden II—æstetisk vurdering og visualisering). Copenhagen, 1998, in Danish.

11. European Commission Directorate General for Energy (DG XVII), European Wind Turbine Catalogue, 1994. Brussels.

12. Danish Technological Institute (Dansk Teknologisk Institut), *Visualizing Wind Turbines in the Landscape* (Visualisering af vindmøller i landskabet), 1994, in Danish.

13. Frode Birk Nielsen, *Wind Turbines and the Landscape: Architecture and Aesthetics* (Aarhus, Denmark: Birk Nielsens Tegnestue, 1997), video. This film was recognized as a finalist at the 3rd International Environmental Film Festival, Pretoria, South Africa, 1997.

14. Frode Birk Nielsen, *Location of Wind Turbines* (Om vindmølles placering, 1. generelt & 2. projekter), School of Architecture (Arkitektskolen), Aarhus, 1980, in Danish. The report appeared in an edited form in Essam El-Hinnawi and Asit K. Biswas, *Renewable Sources of Energy and the Environment* (Dublin: Tycooly International, 1981), Chapter 4, pp. 110–113.

15. Danish Ministry of Energy (Miljö-og Energiministeriet). *Energy-plan of 1981* (Energiplan 81).

16. Data compiled by Energi og Miljeodata in Aalborg, an independent source of statistics on the development of the Danish wind industry. They are also a source for wind farm planning and visualization software.

17. This is also equivalent to 50 percent of the electricity consumed by Denmark's 1.3 million single-family homes, or the electricity consumed by the city of Copenhagen.

18. Danish Energy Agency (Energistyrelsen). *Action Plan for Wind Turbines in Danish Waters* (Havmøllehandlingsplan for de danske farvande). Copenhagen, 1997.

7

LANDSCAPE AND POLICY IN THE NORTH SEA MARSHES

CHRISTOPH SCHWAHN

By the start of the present millennium Germany had installed more wind generating capacity than any country in the world. Nationwide, wind turbines produced nearly 2 percent of the country's electricity. In the state of Schleswig-Holstein, the turbines provided 19 percent of supply. The dramatic growth of wind energy in Germany has occurred within the context of a strong desire to protect the environment. Drawing on surveys in the northern polderlands and experience gained in "reading" different landscapes, Schwahn believes we must be realistic about the use of wind power, that we also must encourage reduction in energy demand, and that in all cases the most effective way to minimize landscape conflicts is to incorporate public views early in the design process, dedicating some areas to wind energy while excluding it from others.

A CONTROVERSIAL POLICY

For many years, the German public has been debating how best to generate electricity. Sometimes the discussion has become heated; it degraded into something resembling a civil war when sites for proposed nuclear power plants, such as at Brokdorf and Grohnde, turned into battlegrounds.[1] Thousands of policemen supported by tanks and helicopters faced thousands of demonstrators. These confrontations over nuclear power were followed by a debate over the destruction of forests due to acid rain from Germany's use of hard coal. Because the production of hard coal has been heavily subsidized to maintain German jobs, the government was not particularly fond of a broad public discussion about whether it makes sense to generate electricity from coal. The debate then shifted toward the personal use of automobiles, focusing on concerns about changes in

Copyright © 2002 by Academic Press.
All rights of reproduction in any form reserved.

global climate, and the policies needed to reduce CO_2 emissions. In Germany, as elsewhere, the public viewed all the most common ways to generate electricity as having significant drawbacks.

Until 1991, wind energy played no important role in German energy policy discussions mainly because there were so few wind turbines. In the context of the ongoing and sometimes violent debate about nuclear power, this seems contradictory. Indeed it is. While other countries such as Denmark began developing and erecting wind turbines in the 1980s, nothing similar took place in Germany. After the failure of a gigantic government-sponsored wind turbine called GROWIAN, few government officials believed wind energy had a future.[2] They held this belief because wind turbines did not fit into the German system of centralized electricity generation. However, when Danish technical advances became evident to the public, the German government reconsidered the wind option. In 1991, Germany's federal parliament passed the so-called "Electricity Feed Law" (*Stromeinspeisungsgesetz*), guaranteeing wind turbine operators payment of 90 percent of the retail price of electricity. In addition, some states (*Länder*) also offered attractive public grants and other subsidies to investors. For example, Schleswig-Holstein's ministry of environment informed farmers that the total investment in a wind turbine could be amortized within 10 years. This information seemed to catalyze radical changes in the landscape, especially along the North Sea coastline.

The differences in public attitude toward wind energy that followed could be called radical. A journalist in northern Germany termed the change in attitude a "gold rush" (*Goldgräberstimmung*), while another created the term "wind rush" (*Windrausch*) to describe what was happening.[3] Such descriptions suggest several attitudes: that wind energy development has more to do with money than with the environment, and that wind generators could be regarded as government-sponsored money-making machines. One thing was clear: despite years of debate about energy resource development, there had been very little effort to develop new energy technologies before the launch of the government's wind energy program.

The new "rush" produced a rapid development of wind turbines (Figure 7.1). Today, a single 1.5-MW turbine is equivalent to 10 of the first-generation turbines and arguably has less impact on the landscape. The planning process itself, which determined just where wind turbines could and could not be erected, was overloaded and differed from one local authority (*Landkreis*) to another. Even subtle connotations of the term *Windrausch* can provide a sense of the atmosphere of wind energy development. For example, *Rauschen* is German for "to rush," which is

FIGURE 7.1   Hessen (Hesse), Mittlegebirge. Two turbines in a mixed cluster of machines on a hilltop in the Hoher Westerwald of Germany's central highlands. The Vestas turbine on the left is in the 500–600 kW class and was made in Denmark. The Fuhrländer on the right is about the same size and was built within the region. The turbines use rotors about 40 meters (130 feet) in diameter and stand atop towers of about the same height. Other turbines may be seen in the distance. (Courtesy Paul Gipe.)

used much the same way as in English to describe the sound of "the rushing wind." However, the substantive of *Rausch* means drunkenness or a state of being "high." *Windrausch*, then, could be interpreted as an intermediate state created by politicians to curry votes in the next election, and perhaps to intoxicate their electors with a short-term program for promoting wind energy.

Not surprisingly, the public discussion which followed the changes in north German landscapes became highly polarized. Even though nuclear power and coal were clearly encumbered with visual and environmental disadvantages, no one could risk objecting to alternative forms of energy in general or oppose particular projects without being overrun by the wind energy lobby or antagonizing certain elements of government.

In the heat of the energy debate the growing conflict between global environmental concern and local landscape protection was ignored, and for a time was even suppressed by some officials. For example, Lower Saxony's minister of environment, Monika Griefahn, withheld the results of a workshop on the placement of wind turbines into the landscape. She also was responsible for revising a land use map of Lower Saxony, which delineated wind energy exclusion zones around seabird breeding reserves, before the map was published. These actions discouraged a frank discussion about the need for landscape protection. Still, some nature-protection societies, such as the Bund für Umwelt und Naturschutz Deutschland (the German Association for the Protection of Nature and the Environment, or BUND), pointed out the need for balance between wind energy development and landscape protection.[4]

Despite all the debate, there has been no substantive change in German policy toward electricity generation. In 1992 the government published a plan for seven new nuclear power plants. Public demands for a revision of electricity pricing to prevent waste by big consumers went unheeded. Although the need to develop renewable energy is constantly touted by officials, such energy development still seems intended to complement, rather than to replace, conventional energy sources. In spite of publicly funded wind energy programs, this energy resource represented only 1 percent of the German electricity supply in 1998, a negligible factor in the total amount of production of electricity in Germany. Although wind's contribution doubled to 2 percent by 2000, no substantial change in German electricity policy is foreseen. It is difficult to go in a different direction without changing general attitudes. The "dogs are barking," as our former chancellor Helmut Kohl used to say, "but the caravan goes on."

A FALSE SOLUTION TO ENVIRONMENTAL GUILT

Characterizing all opposition to wind energy as self-centered responses is an oversimplification. The situation is far more complicated. When people feel personally guilty, they usually do not like to dwell on it. Yet if threatened by change, they consider their options. In Germany, as well as in most Western countries, people often feel guilty about the environmental destruction caused by their affluent lifestyles. In spite of this, there is little movement toward the fundamental changes necessary to reduce our impact on the environment.

This paradox leads to many contradictions. Commercial aviation, for example, has a very poor reputation in Germany. Despite battles between police and opponents over the construction of a new runway at Frankfurt am Main's airport, the number of flights has steadily increased. The reason is simple. More people are flying more often. Consider that in 1997 a member of the Göttingen city council representing the Green Party (*Die Grünen*) boasted incongruously to a local newspaper that by chance he met another Göttingen councillor during their holidays in New Zealand. The happenstance of this chance encounter, of course, was bought with an enormous amount of fossil fuel, a contradiction lost on these members of *Die Grünen*, a self-proclaimed proenvironment political party. To live with such contradictions, people look for easy solutions.

The relationship between environmental guilt and wind power development is evident in public opinion polls. My colleague Jürgen Hasse and I evaluated the visual impact wind turbines might have in the Wesermarsch district of Lower Saxony, a low-lying district (*Landkreis*) northwest of Bremen between the mouth of the Weser river and the Jade Bay. As part of our study, we surveyed tourists about their attitudes toward wind energy, and we found that many of them considered wind power "good because it can replace the use of nuclear power." The results of our survey led us to the conclusion that wind energy is good, because the public wants to believe that the electricity they are consuming is made from wind energy and not from nuclear fuel. Wind power generation allows the public to feel good about their electricity consumption, even though most electricity is still supplied by conventional sources, including nuclear fission.

Environmental guilt has even affected the financial viability of wind power. It is no secret that growth in Germany is not the result of an open competitive market, but is dependent upon subsidies and above-market prices. Power companies have complained that they are forced to offer tariffs to "green" energy producers which, in comparison to conventional sources, seem much too high to them. Court decisions have affirmed the constitutionality of Germany's electricity feed law, at the same time calling for a better solution to the conflict.

In the minds of the local authorities we worked with in the Wesermarsch, the growth of wind energy appeared unstoppable because of the lucrative subsidies available then and the high payments from the electricity feed law, all of which are subject to the whims of politicians. In the mid-1990s, changes to the planning process were proposed to facilitate the siting of wind turbines and wind power plants, but local authorities declined to participate, preferring to wait for political changes in policy.

A comparison between Germany and an isolated spot in Great Britain illustrates how the use of turbines can be connected to energy awareness. On the remote island of Foula in the Shetland Islands, residents became accustomed to having to start a noisy diesel generator before switching on their washing machines. Then the Shetlands' governing council installed a wind turbine and built a small storage reservoir on a hill. Although the landscape of the Shetlands is unique and scenic, no one objected to the wind turbine. Today residents are proud of their new electrical system. They have also become more aware of how much energy they use and try their best to keep consumption of electricity to a minimum. Everyone understands this, and they adjust their consumption accordingly. They defer their washing until it is windy. However, such awareness exists neither in Germany nor in most western countries, where there is no cognitive melding of electricity consumption and the erection of wind turbines.

REASONS FOR LANDSCAPE PROTECTION AND PLANNING

Since land is a resource essential to human activity and life, we have always influenced its character. Most landscapes can be considered as permanently altered memorials to humans and their varied and changing ways. However, it is obvious how much more dramatic the changes have been during the industrial age than in previous periods. The industrial revolution over the past 150 years welcomed new technologies, often overlooking the fact that new technologies frequently come shackled to undesired consequences. In many regions of the world, industry and technology have completely altered the landscape, creating a host of environmental and aesthetic problems. Landscapes are the silent witnesses to these changes.

Yet only humans are able to draw lessons from the past and to anticipate the future. Further, we are the only beings capable of developing ethical standards to regulate our conduct. As a result of the environmental damage produced from exploiting natural resources, our awareness has grown that our activities endanger not only other species on the planet, but ourselves as well. Thus, our understanding of ecological interdependence necessitates landscape planning to prepare responsible and sensible guide-lines to using natural resources.

The need for landscape protection has developed in response to some painful losses. A part of the earth's surface is not only landscape but also *Heimat*, the homeland of the people who live there. People acquire a mental image of their homeland, one which is hardened against the rapid landscape changes that can be brought on by modern technology. They can, in effect, feel expelled from their homeland without ever physically leaving. Unfortunately, this condition has not often been taken seriously, perhaps because it is subjective.

Its subjectivity may also explain why those who are concerned about the impacts of wind power projects don't dare say: "I don't feel at home any longer, because it doesn't look like my home any longer." Instead they argue that birds may be endangered, that the amount of electricity produced is ridiculously small compared to a conventional power plant, or that there is a danger from flying wind turbine blades. It is perhaps instructive to note that in Germany as well as other Western societies, there have been many studies on the impact of wind turbines on animals, birds in particular, but little research on the impact of wind power on people. The most important problem for wind developers is how to overcome the public sense of angst brought about by the rapid changes in the landscape that wind development can bring.

EVALUATION OF TURBINE PLACEMENT IN NORTHERN GERMANY

Placing turbines in the landscape and keeping the public involved in the process are two of the most critical steps in the acceptance of wind power. In that regard, I would like to describe some of the lessons that Jürgen Hasse and I learned through our study of the polder landscape of the Wesermarsch in northern Germany in 1992.[5] Among the most obvious limitations of turbine placement is their verticality and the unavoidable need that they be erected on tall towers at exposed sites. They cannot be hidden behind hills or trees. With the potential for high visibility, care must be taken in the selection of sites, as well as in the design of the turbine and the tower. Design is sometimes made by intuition, but an intuitive decision is only made by one person. Democratic decisions are, by their nature, not individual ones, and so in the decision-making process the relevant criteria for aesthetic design must be known and accepted by the group responsible. This is why systematic evaluation of aesthetic criteria is essential to democratic planning.

In German landscape planning, evaluation is made for a variety of purposes:

- An evaluation of the suitability of a landscape for a special use such as for tourism or for wind energy (*Eignungsbewertung*)
- An evaluation of the impacts on a landscape, because German environmental law requires compensation for impacts on environmental and aesthetic values (*Eingriffsregelung*)
- An evaluation of certain landscapes or parts of them, for example special ecological habitat, to determine the number and kind of flora and fauna, and to determine possible mitigation strategies

LANDSCAPE PERCEPTIONS

Aesthetic assessments face a unique problem because aesthetics are difficult to quantify. Whereas environmental assessments are expected to be scientific—that is, based on objective and measurable criteria—aesthetic perception is entirely subjective. This distinction does not mean, however, that aesthetics are less important or inferior to more measurable criteria. To the same degree that everyone has material needs such as food, energy, and shelter, we all also have nonmaterial needs such as love, identity, and beauty. The importance of different nonmaterial needs varies from person to person; this is true of material needs as well. Similarly, aesthetic values such as beauty, variety, and individuality in a landscape may be appreciated individually, and thus subjectively, but they are no less real.

Although aesthetic characteristics defy quantification, they can be described. For example, one can usefully analyze characteristics of the landscape and how they are perceived. The marshes of Friesland along the North Sea coast are extremely flat. Someone who has not learned to appreciate this special kind of landscape might call it monotonous. Without systematic analysis, I would have been unable to describe differences between various landscape structures in this coastal zone. We were intrigued to find that the landscape units we identified in our analysis corresponded to different epochs of marsh formation.

An analysis of the kind we undertook is simple but requires substantial field work. The first step is a quick tour through the study area to get an overall impression. Sometimes it is necessary to repeat the survey in different seasons, at different times of the day, and under varying weather conditions. Evaluating the impacts wind turbines may have also depends upon changing the viewing distance to provide varying fields of view which are relevant for experiencing a landscape, and the viewshed from which a future turbine can be seen. It is particularly important to differentiate between foreground, middle distance, and distant views. A survey such as ours can help identify criteria to be used that may differ from one type of landscape to another.

Tall structures can clutter and impede the view, dividing landscapes into different vertical spaces. It is useful to identify barriers or obstacles and the spaces in between them, as for example in the marshes of the coastal North Sea where the rate of structuring is important in differentiating landscape spaces and in judging their sensitivity to the visual intrusion of wind turbines. Such structuring can then aid in evaluating landscape harmony, a subjective step but one that can be approached descriptively. For example, one can describe the elements which shape the natural scale of a landscape such as the height of trees, and the elements which pierce it such as television towers, wind turbines, harbor cranes, and tall buildings. In northern Germany, the relationships between vertical and horizontal dimensions are particularly important and markedly different from those in mountainous regions such as Palm Springs, California. This difference can influence public reactions to wind installations. In northern Germany an airplane hangar with a large surface area but little height may produce less aesthetic impact than a wind turbine that requires little surface area but which needs a tall tower.

An important criterion in considering turbine placement in flat landscapes is the line of the horizon, because defining the horizon may not always be easy. How much of the horizon is visible? Is the horizon line sharp or is it blurred? Is it broken by vertical elements? What are its characteristics? Such images can be compared with silhouette lines in mountainous landscapes, although the horizon of a flat landscape plays a bigger role in visual perception than the ridge line in mountain landscapes. In flat landscapes the view changes very slowly as you travel. In the mountains, silhouettes constantly change with regard to the observer's location. There's a saying in North Friesland that illustrates this: "In the marshes you can see today who will come to visit you the day after tomorrow."

Strong vertical elements often landmark flat terrain. In the past, the tallest structures in the marshes were church steeples. Together with lighthouses, church steeples were once used by coastal sailors as navigation markers. Even today, their value as landmarks can be seen in the marshes where there are no hills to climb for orientation and the often cloudy sky obscures the sun. Wind turbines, when erected in such areas, become our newest landmarks. These examples illustrate the value of landscape interpretation: everything in the landscape says something, tells something about the people who live there and something about their relationship with the land.

The age of landscape elements also influences our reactions to them. Most of the older landmarks on the polders, such as traditional windmills, are now surrounded by trees. Trees and shrubs also surround the buildings of historic settlements as a protection against the constant wind. Although put into the landscape by settlers, these plants are today described as natural forms.

Whereas individual observers see objects subjectively, there remain many common interpretations. We see trees as natural elements in a landscape, and we will always view church towers as cultural elements. Harbor cranes, electrical transmission towers, and wind turbines are usually seen as elements of technical civilization. With the passage of time, one day some of these structures too might be seen as cultural elements, as has occurred with traditional windmills. To describe what a landscape is saying, one must identify its cultural and natural elements and interpret its meanings. Once this step is completed, it will be easier to estimate a possible change in landscape expression caused by the addition of new elements.

Technical landscape elements are often standardized and quite similar in appearance. As their number on the landscape increases, their identification as specific landmarks diminishes. The multiplication of standardized elements, such as electrical transmission towers and wind turbines, decreases the ability to orient oneself within the landscape. Formerly, residents of Germany's polderlands could distinguish every church tower and identify the name of the village to which it belongs. Today, wherever you look in the polderlands, you will see the turning rotors of wind turbines. Because of the repetition of these visual elements, wind turbines can be very annoying, contributing to a standardization of the landscape like that caused by industrial agriculture. This observation has special meaning in tourist regions such as the North Sea coast. Breaking the repetition, however, can produce aesthetic conflicts as well. At least at present, developers tend to avoid placing different types of turbines within the same grouping. Although there is a need for more specific studies on the design of wind turbines and wind power plants, including means for creating their own individuality, the most effective way to avoid landscape standardization with wind turbines is to dedicate some areas and to exclude others. This decision is indeed crucial in planning for wind energy.

After completing a landscape analysis and identifying different landscape spaces, it is not difficult to describe those that are sensitive to visual impacts. In the landscape of the Wesermarsch, for example, the older polders are characterized by isolated farms whose buildings are nearly invisible because of the trees surrounding them. Prior to 1991, all electric distribution lines in the area were buried, in part to reduce their visual impact (Figure 7.2). We can describe this kind of landscape as having achieved harmony between its human occupants and nature. Installation of wind turbines in such a landscape disturbs this visual harmony (Figures 7.3 and 7.4).

In contrast to the most established polders, younger polders have been shaped by modern agriculture. There are no hedges, very few trees, and long, large fields—all adaptations for modern farm machinery (Figure 7.5). Even though wind turbines would not be hidden here at all, they would cause less disruption than in the older polders where the turbines rise above the treetops. In an industrial landscape, such as near cranes and other harbor structures, wind turbines cause little disruption. Such evaluations, as part of an aesthetic analysis, are essential for decision makers.

FIGURE 7.2   Old polder in Wesermarsch. The last electricity distribution lines were buried only a few years ago. As a result the landscape is characterized by natural elements like trees and hedges. (Courtesy Christoph Schwahn.)

FIGURE 7.3   Old polder in Wesermarsch. Nowadays, the same wind from which earlier residents sought shelter is being used to generate electricity. (Courtesy Christoph Schwahn.)

WIND ENERGY AND VISIBLE POLICY CHOICES

By virtue of its aesthetic impact, wind energy offers the public an unprecedented opportunity to participate in energy policy decisions. Apart from the energy and resources used in the fabrication, the decommissioning, and the eventual dismantlement of a wind turbine, little about wind energy is out of view. The impacts are visible and often audible. There is no ambiguity about the existence of a wind turbine on the landscape. It is there or it is not. There is also no question about its potential hazard as a large machine. Standing underneath the rotor, you feel and hear the immense power of its turning mass. Seeing, hearing, and feeling all this, you become aware that producing electricity has its price, even when it is made out of wind energy. As Martin Pasqualetti points out elsewhere in this volume, wind generators can teach people how precious electricity is and encourage them to be conscientious about their use of it (Figures 7.6 and 7.7).

Because of the characteristics of wind power, it makes little sense to deny that using wind energy produces impacts. Propagandists for nuclear

FIGURE 7.4   Old polder in Schleswig-Holstein. Similar to early polders in Lower Saxony, the polder in this photo is characterized by trees, hedges, and older buildings. The farm building is built in the traditional style with brick and thatch. Enercon E40 wind turbines are in the background. (Courtesy Christoph Schwahn.)

power, by contrast, have denied its impacts for decades, an easier task because most of the risks and negative effects of nuclear power are invisible and long-term. The risks from wind energy, however, can be anticipated by nearly everyone simply by looking at the turbines. When a risk is known, people can develop policies to compensate for it. Because the impacts and risks of using wind turbines are clearly limited, developing a responsible policy for using wind energy is easier as well.

If wind energy is presented as an alternative to conventional electricity *production*, it should also be presented as an alternative to conventional electricity *consumption*. Germany has more installed wind capacity than any other country in the world. Yet wind energy still accounts for only 2 percent of Germany's total electricity consumption. This illustrates a flaw in German energy policy: it is not sufficient just to develop new technologies; they should be employed in an intelligent and environmentally sensitive way as well. There should be an overall plan for incorporating them into the economic infrastructure. Unfortunately, Germany seems to be a long way from institutionalizing this approach.

FIGURE 7.5   Newer polder in Schleswig-Holstein. Like more recent polders in Lower Saxony, these modern polders are empty and speak of a more industrial agriculture. The tanks treat and store liquid manure. The low ridge in the foreground is an old dike. (Courtesy Christoph Schwahn.)

Wind energy is presently plugged into a system which was designed decades ago when environmental concerns were less of a priority. Twenty years ago, Lower Saxony's prime minister Ernst Albrecht predicted that by the year 2000 nearly 50 percent of the state's homes would be heated electrically. At the time planners expected this electric heat would be provided by large nuclear power stations. Of course with the shift away from nuclear power, this will never be achieved. Still, the general structure of the electric utility system has not changed. Electricity is still produced in the traditional, centralized way. Pricing discriminates against small consumers by offering discounts to large customers. And although even many children know that two-thirds of primary energy is lost in conventional thermal power plants, heating with electricity remains fashionable because consumers find it more convenient. The question then is identifying not merely the source of the energy, but what it is being used for.

To the adage "time is money" can be added "speed is energy." Increasing the pace of society increases our consumption of energy. The similarity between spinning wind turbine rotors and the wheels of a high-speed train should make us think about our transportation choices, too. For

FIGURE 7.6   Lower Saxony. Northern Germany. Billboard in the sky. Early Tacke 80-kW turbine at a truck stop along an autobahn near Salzbergen, Germany. Some promoters in China and Eastern Europe have used the entire tower to advertise various companies or their products. (Courtesy Paul Gipe.)

example, the German magazine *Der Spiegel* noted that 16 wind turbines of 600 kW each would be necessary to supply one Intercity Express (ICE) train with electricity.[6] It would be ironic indeed if the energy used to satisfy our desire for high-speed convenience comes from wind energy, given our abandonment in the 19th century of an age-old wind-powered technology used for travel: sails.

It might appear naive and idealistic to expect wind energy's proponents to demand a reduction in electricity consumption at the same time they demand expansion of their technology. However, we should recall nuclear proponents' irresponsible promotion of their technology and the conse-

FIGURE 7.7 Lower Saxony, Northern Germany. Enercon is one of Germany's leading manufacturer's of 1.5-MW wind turbines. Here is one of Enercon's E66s outfitted in anticollision markings looming over Enercon's blade assembly hall in an industrial suburb of Aurich in Östfriesland province. The E66 uses a rotor 66 meters (215 feet) in diameter on a tower of equivalent height. (Courtesy Paul Gipe.)

quences this had for public perceptions. We must note that if wind energy is not to become just a supplement to conventional sources but a true alternative, then it must be employed in a truly alternative way. Proponents of wind energy are held to a higher standard than those of the nuclear industry largely because of the public's disillusionment with the promotion of nuclear power.

Wind energy advocates should make clear that it would be difficult to satisfy our present need for electricity solely with the source they prefer.

We have to confront our guilt about exploiting the planet and avoid being misled into believing that new technologies alone will solve our problems. Only by confronting our so-called *need* for electricity can we develop a more responsible policy for the *use* of energy, truly assuaging our complicity in plundering the planet.

Fundamentally, two ways exist to increase the share of wind energy in the supply mix: to install ever more turbines in the country or to reduce overall electricity consumption. Energy policies of the future will surely involve a combination of the two. In Germany, as elsewhere, the latter approach promises as much if not more progress toward that goal than simply installing more wind turbines.

Advocates should not be single minded. Wind energy is not the only important form of renewable energy. There are others as well, such as direct solar energy. Again, this illustrates that wind energy must be a part of a new energy policy in which all forms of renewable energy will play a role where they are best suited.

Producing energy is not the only demand on public resources. We have to respect the aesthetic desires of the people, and we must try our best to insert wind turbines into the landscape in a responsible, thoughtful way. We have to acknowledge the needs and fears of the people affected, because those who profit from electricity generation are typically not the people who suffer from its production. Cities such as Berlin, for example, could contribute far more to an alternative energy system by reducing their consumption than by installing a few wind turbines inside the city. Producing energy is not an end in itself.

NOTES AND REFERENCES

1. In February 1981, 100,000 opponents of nuclear power faced 10,000 heavily armed police outside the village of Brokdorf, population 800. Police used water cannons and tear gas against the demonstrators.
2. For more on GROWIAN see Paul Gipe, *Wind Energy Comes of Age* (New York: John Wiley & Sons, 1995); 105–108.
3. Community of Reussenkoege Intends to Plan Wind Farm (Reußenköge wollen Windpark ausweisen), in *Nordfriesische Nachrichten*, January 5, 1991; and Windmills Only for the Locals? (Windmühlen nur für Einheimische?), in *Nordfriesische Nachrichten*, February 26, 1991, Niebuell, Germany.
4. BUND is Germany's largest environmental organization. An affiliate of Friends of the Earth with 300,000 members, BUND should not be confused with BLS, Bundesverband Landschaftschutz (the German Association for Landscape Protection), an anti-wind group much like Country Guardians in Great Britain. The BUND chapters in Baden-

Würtemberg, Bavaria, and Thuringia have gone so far as to help sell shares in a cooperative wind turbine.

5. Jürgen Hasse and Christoph Schwahn, Wind Energy and Landscape Aesthetics: Example Wesermarsh. Interdisciplinary Study in three Parts, ordered by the County of Wesermarsch (Windenergie und Ästhetik der Landschaft: Beispiel Wesermarsch, Interdisziplinäre Studie in drei Teilen, im Auftrag des Landkreises Wesermarsch), 1992, Göttingen und Bunderhee, Germany, in German.

6. Wind Energy: The "Tuddlemast" Parade (Windenergie: Parade der Tüddelmasten), *Der Spiegel*, 39/1995, pp. 194–200, Hamburg, Germany, in German. ICE trains are the German equivalent of France's TGV or Japan's bullet trains.

WORKING WITH THE WIND

8

LIVING WITH WIND POWER

IN A HOSTILE LANDSCAPE

MARTIN J. PASQUALETTI

*At the same time wind development was taking shape
in Altamont Pass, thousands of wind generators sprouted
with surprising speed from the harsh desert near Palm
Springs 400 miles to the southeast. The unfamiliar
devices were soon generating more controversy than
electricity, and everyone from entrepreneurs to politicians became part of the debate. San Gorgonio Pass
today has matured as a site for wind power, the experience there amounting to a landscape laboratory where
there has been a softening of opposition and an accom-
modation of sorts between the landscape that was and
the landscape that is.*

The broad high-level participation at the 1992 Earth Summit in Brazil
and the 1997 Kyoto meeting on climate change was a sign of rising global
concern for the health of the planet. At both meetings, as well as at
hundreds of smaller meetings since, one of the most important questions
has been how to reduce the environmental price of energy demand. One
response has been to promote improvements in energy efficiency.
Although programs of this type have successfully reduced both demand
and pollution, such one-time improvements are soon overwhelmed by
greater energy demands produced by greater numbers of people and
improved lifestyles.

All the attention that the clashes between energy and environment have
received has served to educate the public to a greater degree regarding the
various connections between the two, but it has not been an easy task: it
has had to erase centuries of experience that told us that we could have
little impact on nature, either because natural systems were so huge that
they could not be damaged, or because we were too weak and puny to do

Copyright © 2002 by Academic Press.
All rights of reproduction in any form reserved.

anything to restore what damage did occur. Either way, we tended to dismiss the growing evidence of wounded landscapes.

Today we have to admit that the bliss of ignorance has substantial dangers. With world population growing at 80 million souls each year and energy demand rising at an even faster pace, we now have the power to overwhelm every natural buffer that is built into the biosphere. Threats there were once on the margins are now in position dead ahead. As part of a portfolio of responses we have been trying to promote energy resources less threatening to our finite Earth, the only home we have.

We now accept that we face a serious challenge. The energy we would like produces no waste, dirties no skies, dams no rivers, floods no canyons, poses no lingering threats to future generations, all the while remaining unending and affordable. Does such a resource exist? The answer is yes. With a bit of good technological and economic timing, our requirements can be partially met by a resource familiar to us all, one positioned by history, research, development, capacity, and economics to be of significant near-term help. I refer here to the ubiquitous and unending power of the wind.

A PARADOX OF POWER

Although wind power produces electricity by a process that is clean, affordable, and available, one cannot easily dismiss the fact that in many places it has received an unexpectedly chilly reception from the public. What is the explanation for this reaction? How has the benign environmental reputation of wind power fallen on such hard times? What does this turn of events suggest for the future renewable energy resources that we had hoped would keep the environmental noose from tightening around our necks? And where did this hostility originate? One place to look for answers is in southern California.

Although wind has been used for many centuries to propel ships, grind grain, and pump water, its use to generate electricity is more recent, beginning in earnest in the mid-1980s in three areas of California, including the San Gorgonio Pass near Palm Springs, 100 miles east of Los Angeles (Figures 8.1 and 8.2). From the outset, the development of wind power near Palm Springs has been not only conspicuous and controversial, but even suspect. The erection of wind turbines on a patch of land long considered of no commercial value was so unpopular that it led quickly to legal responses, political battles, regulatory sanctions, and even a smattering of public loathing.

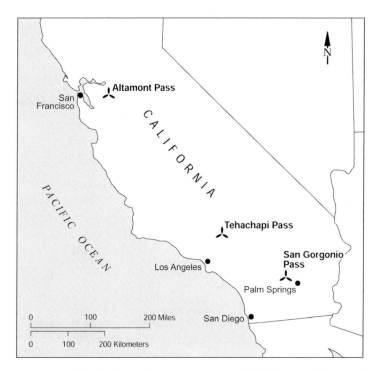

FIGURE 8.1   Principal wind resource areas of California. (Cartography by
Barbara Trapido-Lurie, Department of Geography, Arizona State University.)

The experience has been instructive at several levels. The wind power
industry, expecting a more cordial welcome, realized that many adjustments in strategy, deployment, and engineering were going to be necessary
if wind was to succeed here. Indeed, the entire alternative energy industry
soon learned a lesson: do not take public support for alternative energy for
granted, even in progressive California.

The harsh San Gorgonio Pass experience was not an isolated public
response, but it was among the most noteworthy. In other states, and
especially in Europe, the public has reacted with similar skepticism to
wind developments. In England, wind projects are at a standstill, pending
ongoing debate about aesthetics. Wherever such complaints have been
recorded, it is usually possible to trace their origins to California. It was at
San Gorgonio Pass that wind promoters first realized that tapping the wind
would not escape scrutiny or criticism.

Much of the attention wind power receives, both positive and negative,
emanates from its intrinsic spatial contradictions. No other energy landscape is simultaneously so intrusive yet benign, so dynamic yet site-

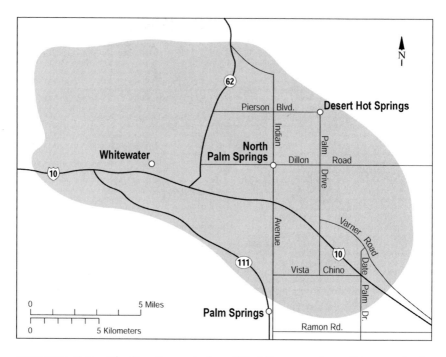

FIGURE 8.2   The San Gorgonio Pass Wind Resource Area. (Cartography by Barbara Trapido-Lurie, Department of Geography, Arizona State University.)

specific, so hated by some yet championed by others, so chaotically distributed in one place while being neatly regimented in another. Moreover, unlike many alternative energy choices that are being promoted, it competes economically with conventional sources. In many ways, wind energy is a paradox of power.

A NOTCH IN THE MOUNTAINS

San Gorgonio Pass is a constriction between Mt. San Jacinto to the south and Mt. San Gorgonio to the north (Figure 8.3). Strong and consistent winds have whistled through the pass for centuries, bending plants, polishing rocks, and piling up sand into large dunes.[1] Archaeologists tell us that it was part of a route used for centuries by those trekking between the desert and the Pacific. Native Americans even developed a local legend: "When the wind quits in the pass, the end of the world will have come."[2]

Many threads vital to the Los Angeles infrastructure journey through this important notch, including railroads, Interstate 10, telephone cables

FIGURE 8.3   Oblique photograph of the topographic constriction of the San Gorgonio Pass. Mt. San Gorgonio is on the right (north) and Mt. San Jacinto is to the left. Northernmost Palm Springs is visible at the base of Mt. San Jacinto. (Courtesy Martin Pasqualetti.)

and fiber optics, electrical transmission lines, aqueducts, and petroleum pipelines. Even smog blows through the pass almost every day in the summer, all courtesy of the San Andreas Fault, which helped create this great cleft in the first place.

When 20th century settlement began in the desert, strong winds were inescapable, and nothing has changed. The winds sometimes still topple road signs and overturn trucks. Blowing sand still pits glass, strips paint off cars, and even severs exposed telephone poles near their bases. Sentinel rows of salt cedar and eucalyptus are used to shield crops, and sand fences protect houses and cars. With homes priced higher in the calmer southern part of Palm Springs, wind even figures into real estate values. The wind has always been part of life in the desert.[3]

What has changed, of course, is that now the wind is visible. When the first modern turbines were constructed in the mid-1980s, they produced a cascade of complaints about their unsightliness, the noise they produced, the birds they threatened, the potential danger they represented from structural failure, the hazard they posed to aircraft, and the electrical interference with television reception. In the beginning, no one seemed to like the wind turbines regardless of where they were placed. This caught the industry unawares: their siting philosophy had seemed foolproof because the windiest sites were notably barren of competitive use and lacked any local sensitivity.[4]

LANDSCAPE LABORATORY

Objections to San Gorgonio Pass wind development began as soon as the wind turbines started rising quickly from the sand. The news spread quickly. It was a big event. Living nearby, I often overheard people commenting that the wind turbines were "ruining the desert." The local newspapers carried complaints and political condemnations. As a geographer, I was fascinated by how quickly and completely the wind turbine installations transformed a desolate patch of real estate into an evocative landscape of power. I wondered if the San Gorgonio Pass experience would be repeated in other locations.

As with hydropower and geothermal energy, the development of wind power is knitted into the local land use. Although the eastern end of San Gorgonio Pass is sparsely inhabited, it is still a busy place. One of its oldest functions is to provide the principal corridor to the oasis resorts such as Palm Springs, a city with a flamboyant history. After the advent of sound, as the Hollywood film industry took solid form, Palm Springs became a mecca for stars with cars who often took their leisure in this quiet, small, warm, and exclusively isolated desert town. With modern freeways, memories of the long drive of 60 years ago have been lost as waves of visitors can now drive there in two high-speed hours. Thousands more winter residents from places such as Chicago, Seattle, and Vancouver come to the desert seeking to avoid the cold and damp of their hometowns by staying in their desert homes from October to May. The community, known at first as nothing more than a sun-blistered desert hamlet, became a land of luxury resorts, expensive restaurants, and golf courses in such numbers that an enthusiast could play every other day for the entire winter season without stepping up to the same tee twice.

Aside from the turbines, much about the physical landscape of the desert has changed little in the past 50 years. The weather and the topography are much as they have always been. Even the city of Palm Springs itself has been relatively stable in appearance, with most of the new housing and golf courses emerging not in the city itself but further to the south. Especially near the eastern end of the pass, there was a sense of

FIGURE 8.4   Wind turbines in the San Gorgonio Pass, looking northwest toward the San Bernardino Mountains in the mid-1990s. (Courtesy Martin Pasqualetti.)

landscape permanence. As the wind turbines took root, everything quickly changed. They became the dominant landscape feature at the entry point to the Palm Springs area (Figure 8.4).

Today with more than 3000 wind turbines straddling the interstate highway and climbing the mountain slopes, they have become part of the new landscape. In an ironic twist, members of the film industry who once sought the solace of the vacant landscapes to escape the intensity of life in Hollywood are today lured by wind landscapes to use them as stark backdrops for their films and advertisements.[5]

This rich history of notoriety, visibility, and public reaction has made San Gorgonio Pass a landscape laboratory for the study of wind power. Like the attempts at energy deregulation that began plaguing California near the end of 2000, the experience of wind development in California has provided lessons for others as well.

A LANDSCAPE CHANGED

The beginning of the transformation of San Gorgonio Pass into an industrial landscape seems rather recent, but it did have some precedents.

Electricity was first generated from the wind in the 1920s when Los Angeles real estate developer Dew Oliver constructed a 10-ton "blunderbuss." This device, looking like a modern-day jet engine with its narrow midsection, compressed the air by a factor of 12 and actually worked as designed.[6] It was ultimately abandoned when Oliver ran into legal and financial troubles. Fifty years later Southern California Edison erected a single large experimental turbine in North Palm Springs, ostensibly to gather data, but it too was removed after a few years.

More recently, two pieces of legislation changed everything. In 1978, Congress passed the Public Utilities Regulatory Policy Act (PURPA), providing premium rates for renewable energy projects and requiring local utility companies to buy all the electricity that was generated by alternative energy sources. Later, state and federal tax credits created added incentives. Everything was now in place: wind data, cheap and available land, tax incentives, technical expertise, and a guaranteed market. Between 1984 and 1985, wind power took off.

Although the wind was a familiar element of desert life, using it to generate electricity transformed the San Gorgonio Pass and shocked the nearby communities. The blank canvas that had always been there suddenly became an industrial landscape, stunning long-time residents and visitors who had come to expect the desert to forever remain unchanged. As the new additions dominated the landscape, complaints started pouring in from all quarters. People were outraged. Not only had the turbines changed the desert, many of them never even turned, and some had toppled over or lost blades. Furthermore, the noise they produced disturbed the sleep of nearby residents who had built isolated houses in the pass not only to serve their reclusive bent but because they assumed that in such a windswept haven, nothing would disturb their solitude. Once in place, the turbines were often condemned as a "tax dodge for the rich."

Such an accusation, not surprisingly, attracted the attention of politicians. Legal action soon followed, led by the city of Palm Springs. The city sued the U.S. Department of the Interior, claiming that "its U.S. Bureau of Reclamation and the developers had ignored mitigation procedures stipulated in the environmental impact statements, that many of the turbines were non-functioning and were an eyesore, that the inconsistency of sizes and shapes cluttered the landscape, and that the developments threatened the visitor's aesthetic experience and the city's tourism potential."[7] In response to the notoriety that followed this suit and to the many complaints that had been logged by citizens and visitors, Riverside County held hearings, financed a public opinion survey,[8] and created a wind planning document that all future developments would have to follow.[9] This attention would influence wind research and development around the world.

THE RESPONSE

Once it recovered from the unexpected vigor of public resistance, the wind industry responded with a series of initiatives. Trying to educate and sway the public, it organized wind fairs, gave tours, and pointed out that the impacts of fossil-fired plants are much more substantial, far-reaching, and permanent. It noted that generating electricity from the wind produces no toxic waste, no radiation, no acid rain, and no greenhouse gases. All this was of course, true, but no amount of increased public education and understanding could make wind turbines invisible.

It was at this point that governments stepped in, putting in place legal controls, protections, and conditions. Currently, Palm Springs planning ordinances permit wind turbines in zones W, O-5, E-1, and M-2: that is, watercourse zones, open land zones, energy industrial zones, and manufacturing zones. The ordinance specifies safety and scenic separations ("setbacks"), underground collector cables, neutral "environmental" paint color, a 200-foot height limit, advance drawings or photographs of proposed windmills, and a bond for decommissioning in the event of inoperable or dangerous equipment. Outside the city limits, Riverside County imposes similar requirements. More recent additions include legal protections for rare, endangered, and charismatic birds such as eagles and hawks, and both city and county ordinances require filed reports for any bird killed by a wind turbine.

With ambient noise levels in the desert usually much lower than in urban environments, some of the greatest detail in the local ordinances is reserved for noise control. The county, for example, stipulates that noise levels of a Wind Energy Conversion System (WECS) will be 45 dB(A) or lower, unless the noise is considered a "pure tone." All land parcels in the vicinity of a wind farm project used for residential, hospital, school, library, or nursing-home purposes must be identified. A commercial array must be operated at a noise level not to exceed 65 dB(A). It must operate with no impulsive sound below 20 Hz. All noise measurements and noise projections must be made in accordance with the technical specifications and criteria developed by the county health department and adopted by resolution of the county board of supervisors. A toll-free telephone number must be maintained for each commercial WECS project and it must be distributed to surrounding property owners to facilitate the reporting of noise irregularities and equipment malfunctions.

Turbines produce various types of noise, ranging from constant high-pitch frequencies to low periodic pulses. The degree to which these noises disturb people varies with the individual and distance, buffering, ambient noise levels, and turbine design. Noise ordinances passed by city and county governing bodies encouraged technological improvements. Aerodynamic refinements, substitution of tubular for lattice towers, and the switch to the three-blade, upwind designs all made the turbines quieter.

All these measures, plus a degree of resignation and familiarity, have reduced complaints near Palm Springs. However, they picked up in other countries when wind projects were announced there. In the United Kingdom, for example, opponents of wind power campaigned passionately and effectively, bringing development as of early 2001 to a standstill.[10] Yet here and elsewhere, many of the technical and aesthetic improvements that were developed in response to criticisms leveled in San Gorgonio Pass have been adopted elsewhere, sapping the strength of negative reactions.

THE EVOLVING PUBLIC PERCEPTION OF WIND LANDSCAPES

The perception that the nearby wind farms would diminish the attractiveness of the desert as a resort destination led to the establishment of guidelines to govern wind development there, and it could be argued that such guidelines were needed to prevent abuse and protect the citizens. However, a public opinion survey commissioned by Riverside County in 1985 yielded the unexpected finding that, despite the publicity, the public was relatively disinterested in the wind developments.[11] As it turned out, the relatively sanguine public opinions about wind development in San Gorgonio Pass were not unique to that location. Landscape architect Robert Thayer and his associates, for example, reported similar reactions to wind developments on the wetter and gentler topography of Altamont Pass, 50 miles east of San Francisco. And in England, "before and after" surveys by the government's Energy Technology Support Unit (ETSU) in Cornwall in the early 1990s found that wind power was "popular." In addition, the existence of wind farms in southwest England "altered attitudes in the direction of local residents being more favourable toward wind energy," with many of the worries that local residents had about wind turbines having been proved "unfounded."[12] To judge from these and several other opinion surveys of public attitudes toward wind

ACCEPTANCE OF WIND ENERGY

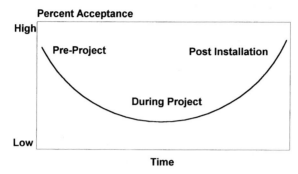

FIGURE 8.5   Sequence of public acceptance of wind power. (Produced by L. Arkesteijn after research conducted by Maarten Wolsink.)[13]

power, public reaction has been following a pattern as identified by Maarten Wolsink in the Netherlands (Figure 8.5).

THE CULTURAL LANDSCAPE OF WIND

Turning the steady winds of San Gorgonio Pass into a steady flow of electrons for southern California consumers had a rocky start, but in the past 15 years it has evolved into a smoother operation. Although wind power still has its dissenters, city and country officials report that public objections to wind power nowadays are virtually "nonexistent."[14] Indeed, acceptance has been on the rise. It is not difficult to find homes sitting squarely within concentrated wind projects (Figure 8.6). Nor is it a rarity to see captivated travelers stopped on the freeway at their own peril to photograph the wind landscapes that dominate the scene. To appreciate how unusual this is, try to recall seeing anyone photographing any other piece of the electrical supply system. Or, try to envision an advertisement which uses images within the coal or nuclear fuel chains to attract the attention of readers as Compaq computers did with a full-page photo of the Altamont wind farms (Figure 8.7). So evocative are wind landscapes that they show up in films, postcards, and even in art exhibits.[15] Clearly, people find modern wind developments unusual, intriguing, and largely a nonthreatening landscape addition, perhaps even a nostalgic reminder of the rustic past when tens of thousands of windmills dotted rural America.

Wind landscapes also mean earnings; if wind turbines can make electricity, they can also make money. This prospect is important not only to investors, but also for easing political opposition, as it did in Palm Springs. A few years ago, the late mayor of Palm Springs, entertainer

FIGURE 8.6   Proximity of wind developments, power lines, houses, and scenery in North Palm Springs, an unincorporated area. (Courtesy Martin Pasqualetti.)

Sonny Bono, was critical of the wind turbines on the north side of his city. But when he learned that they could produce local employment opportunities and tax revenues, he reversed his earlier position and led the effort to sweep an additional 20 square miles of adjacent desert lands into the city's tax base.[16] Although the financial rewards did not match the predictions, the action did have a salubrious effect on wind power by muting political opposition to plans for expansion.

Further support emerged once landowners came to understand that wind projects could enhance rather than diminish land values. They can have this effect because they allow a greater multipurpose and multi-income use of the land. Wind turbines do not require elaborate, expensive, or hazardous infrastructures for fuel supply, power plant construction, emission control, or waste disposal. In Altamont Pass, for example, the added value of wind projects has kept ranch lands out of the reach of housing developers. In San Gorgonio Pass where land is not valuable for agriculture, other types of concurrent use are apparent, including housing, transportation, and recreation. In ways other than aesthetics, wind power places a relatively light and temporary touch on the land. This can translate into profits for the landowner and support for alternative energy development.

FIGURE 8.7   The use of wind turbines in advertising, such as in this Italian newspaper in July 1998, suggests that the public is attracted more than repulsed by wind turbines in the landscape. They are evocative enough for advertisers to use them as props to help sell their products. (Used with permission of Compaq Computer Corporation.)

As a further example of how public opinion has shifted, wind energy is becoming a bit of a tourist attraction. From the mid-1990s, a company has been conducting regular wind energy tours and maintaining a gift shop of wind items for sale in north Palm Springs. In its first 6 months of operation the tour attracted 10,000 people[17] (Figure 8.8). In Tehachapi the local Chamber of Commerce helps organize an annual Wind Energy Fair to promote the contributions of the industry to the community.[18] In Altamont Pass, developers publish brochures and provide tours. In England, where reports of "wind tourism" are even more impressive, the development at Delabole attracted nearly 100,000 visitors in its first

FIGURE 8.8   Wind tours are a popular tourist activity among the wind farms of San Gorgonio Pass. This photograph was taken looking north, near the intersection of Indian Avenue and Interstate 10 in North Palm Springs. (Courtesy Martin Pasqualetti.)

year. These examples are part of a positive scent that is beginning to emanate from wind developments.

Given the rough treatment wind power often receives, what should we make of signs that suggest that many people find wind turbine landscapes increasingly acceptable? It could mean that planning controls and accumulated experience have been effective in encouraging better projects, or it could mean that the public has come to appreciate wind's comparative environmental advantage vis-à-vis more conventional sources of power. Or, is it that many countries have become desperate for alternative energy supplies? Or, have wind projects simply become too profitable to ignore? Although all these factors are part of the explanation, one thing is clear: We find these productive landscapes fascinating. We are attracted to them; we cannot ignore them.

A REVERSIBLE LANDSCAPE

Dams, mines, and nuclear waste sites have a major drawback, their lasting landscape presence. Knowing that wind energy need not carry this burden, the "graveyards" of idle wind turbines in the early days of wind development of San Gorgonio Pass prompted the city of Palm Springs to mandate in their wind ordinance that "any unsafe, inoperable, or abandoned WECS or WECS for which the permit has expired shall be removed by the owner or brought into compliance."[19] The ordinance further stipulates that once a site is cleared, it shall be restored to conditions prior to installation. Bonds are required to cover the cost of removal and site restoration.

Although many idle installations have already been removed by ordinance in San Gorgonio Pass,[20] not all inoperable turbines are restored. As Gipe has pointed out,[21] such removal requirements are not uniform in the United States, although the legal reclamation provisions found in San Gorgonio Pass do have equivalents abroad. In the United Kingdom, for example, "when a wind farm reaches the end of its design life, the turbines can be easily removed and the foundations could be re-used for the installation of new turbines subject to planning permission or, if required, the land could be reinstated."[22] Despite the diffuse nature of wind power and the large number of turbines that are required for a given amount of electricity, decommissioning and removal of the turbines is not a technically difficult or dangerous job. For this reason, wind power need not produce a lasting landscape legacy. This positive trait, in a world increasingly crowded by derelict and redundant industrial equipment, is one of the most conspicuous environmental advantages of wind energy.

THE FUTURE OF WIND POWER

Humans have known the wind ever since it rustled leaves thousands of years ago, whipped up whitecaps on the open sea, and blocked the sun with clouds of dust. Thousands of years before anyone mined coal and uranium or pumped up oil and gas, wind was used to grind grain, lift water, and push boats to new ports of call. Windmills were on the landscape before pyramids rose along the Nile, before Marco Polo crossed Asia, before Columbus reached the New World. Today, we treasure the nostalgia of "tall ships" and place our windmills in museums (Figures 8.9 and 8.10). We ride the wind in sailplanes, surf the waves that winds create, and use electric fans to create our own personal breeze. We even feel oddly out of sorts on a windless day, and sometimes we buy whimsical windmills to give ourselves good cheer (Figure 8.11).

Despite our familiarity with using wind power to amplify human muscle, using it to produce electricity caught us off guard. We felt

FIGURE 8.9 Windmills at the American Wind Power Center, Lubbock, Texas, attest to nostalgia for farm windmills that once dotted the Great Plains in the tens of thousands. (Courtesy Martin Pasqualetti.)

uncomfortable when the quaint and historic wind landscapes of our memories gave way to the similar, yet strangely sterile and odious, wind farms of California. We were confronted with an industrialized landscape which we were simply expected to accept, and after years of schooling about the environmental benefits of alternative energy resources, we were handed the harsh reality such a commitment entailed. The question we yet face is whether we will come to accept wind landscapes, even temporarily, or whether we will abandon them after a short history.

One option, an option I believe we should exercise, is to move along the continuum of wind power, extending it into the future just as it stretches into the past. We benefit from the presence of wind turbines in our backyards because they remind us that our electricity has a cost, that it comes from somewhere. Wind turbines help us appreciate that our energy demand has a price, and that someone must pay it. As wind power expands, we will come to appreciate more fully the advantages that this form of generation promises over other sources: that it poisons no trees, heats no air, triggers no cancers, drowns no canyons, and kills no seals.

Our rich and fruitful past association with wind power would seem to promise a partnership that will be helpful in shaping a hopeful energy

FIGURE 8.10 Windmills at the Shattuck Windmill Museum and Park, Shattuck, Oklahoma. (Courtesy Martin Pasqualetti.)

future. The obstacle that clouds that vision, however, is the shrinking amount of open space at our disposal. As such space becomes an ever more cherished commodity, wind developments will continue to be controversial.

The success of wind power depends on how well the wind industry learns to incorporate the public into decisions, both for the opportunities this allows for broader dissemination of information about wind power and for the suggestions the public can bring to the discussion about their concerns and how to accommodate them. Among the things the wind industry must do is to minimize intrusion, especially in favored places,

FIGURE 8.11 These wind toys for sale at close to Grand Coulee Dam at Electric City, Washington, suggest that the public familiarity and acceptability of wind devices. One would not expect to find similar displays of coal or nuclear power plants. (Courtesy Martin Pasqualetti.)

regardless of the technical attractions such locations may offer. They must also continue to refine turbine efficiency and design, improve spacing strategies and noise suppression, protect wildlife, and practice clean site maintenance and restoration when turbines are decommissioned.

More than any other source of energy, wind power is tied to the land. And more than any other place, the initiation of the modern era of wind development is linked to California. When future archaeologists and historians study the early 21st century, they will note landscape changes that wind power produced, the responses that these changes evoked from the public, and the final contribution that wind power made. One of the richest sites for such research will be the San Gorgonio Pass and the desert oasis of Palm Springs nearby.

NOTES AND REFERENCES

1. Jeffrey A. Lee, 1990, "The Effect of Desert Shrubs on Shear Stress from the Wind: An Exploratory Study." Ph.D. dissertation, Arizona State University.

2. Paul W. Travis, "The Wind Machine in the Pass," *Westways* 48 (February 1956): 12(12–13).

3. According to Battelle's Pacific Northwest Laboratory's *Wind Energy Resources Atlas* (1986), the San Gorgonio Pass resource is 6.3 meters per second (14 mph), yielding 365 watts per square meter. It is a Class 6 area (of 7 classes) at a 10 meter height above the ground.

4. The power in a moving mass of air is proportional to the cross-sectional area of the air mass and the cube of the wind speed. Doubling the radius of a turbine rotor increases the area by the square of the radius, a factor of 4, a direct consequence of the formula for the area of a circle, $A = \pi r^2$. Doubling the wind speed, however, increases the power eightfold. Working with these characteristics, all other things being equal, the wind developer will seek to place the largest wind turbine on the windiest site available.

5. For example, the film *Rain Man*.

6. Robert W. Righter, *Wind Energy in America: A History* (Norman, Okhahoma: University of Oklahoma Press, 1996).

7. *City of Palm Springs* et al. *v. U.S. Department of the Interior* et al. No. CV 85-2004-WMB (Tx), Central District of California, U.S. District Court, entered August 26, 1985. Quotation from Martin J. Pasqualetti and Edgar Butler. Published as "Public reaction to wind development in California," *International Journal of Ambient Energy*, 1987, 8(3): 83–90.

8. Conducted by Martin J. Pasqualetti and Edgar Butler. Published as "Public reaction to wind development in California," *International Journal of Ambient Energy*, 1987, 8(3): 83–90.

9. City of Palm Springs, Ordinance 1472, and Conditional Use Permit 9402.00H8, pp. 263.1–263.10.

10. Interviews with Richard E. Patenaude, Planning Manager, Department of Planning and Building, City of Palm Springs; Paul F. Clark, Senior Planner, Riverside County Planning Department, Indio, California, May 1998.

11. Martin J. Pasqualetti and Edgar Butler, "Public reaction to wind development in California," 1987, *International Journal of Ambient Energy*, 8(3): 83–90.

12. David Elliott, *Energy, Society and Environment* (London: Routledge, 1997): 159.

13. Produced by L. Arkesteijn after research conducted by Maarten Wolsink. "The Siting Problem: Wind Power as a Social Dilemma," paper presented on the European Community Wind Energy conference, Madrid, Spain, 10–14 September, 1990; and Maarten Wolsink, "Attitudes and Expectancies about Wind Turbines and Wind Farms," *Wind Engineering*, 13:4 (1989), 196–206.

14. Richard E. Patenaude, Planning Manager, Department of Planning and Building, City of Palm Springs; Paul F. Clark, Senior Planner, Riverside County Planning Department, Indio, California.

15. For example, the photograph by Nikolay Zurek displayed as part of the exhibition "Picturing California: A Century of Photographic Genius," Oakland Museum, 1987.

16. Personal interviews with Richard E. Patenaude, Planning Manager, Department of Planning and Building, City of Palm Springs; Paul F. Clark, Senior Planner, Riverside County Planning Department, Indio, California, May 1998. Also, see Righter, *Wind Energy in America*, 229–232.

17. Contact through e-mail at Windtour@aol.com.

18. Largely through the efforts of wind power activist Paul Gipe.

19. City of Palm Springs, Ordinance 1472, and Conditional Use Permit 9402.00H8, pp. 263.1–263.10.

20. The ordinance states that "... every unsafe or inoperable commercial WECS and every commercial WECS which has not generated power for 12 consecutive months is hereby declared to be a public nuisance which shall be abated by repair, rehabilitation, demolition or removal (unless the owner can demonstrate that modernization, rebuilding or repairs are in progress or planned...)."

21. Interview with Paul Gipe, May 1998.

22. DOE and Welsh Office, *The Countryside and the Rural Economy*, Planning Policy Guidance Note 7 (London and Cardiff: HMSO, 1992).

9

DESIGN AS IF PEOPLE MATTER: AESTHETIC GUIDELINES FOR A WIND POWER FUTURE

PAUL GIPE

In the course of more than 25 years of observing, speaking, promoting and writing about wind power, Paul Gipe has accumulated a detailed appreciation for the important role aesthetics will play in its public acceptance. He believes that the wind industry challenges its own credibility as an alternative energy source when it does not follow best environmental and operational practice. Here, he proposes almost three dozen specific guidelines for the industry to follow if wind power's contribution is ever to match its promise.

California is a land of dreams. At least it was for me, when I moved to Tehachapi in the spring of 1984. I planned to live the dream. Wind energy had become a reality on the windswept hills east of town, and I was going to help it grow. For those of us who had cut our political teeth in the environmental movement of the early 1970s—in my case the 7-year effort to regulate strip mining—wind energy offered another way, the "soft path," as Amory Lovins calls it.[1]

Yes, wind energy was an alternative to the secrets hidden behind the concrete and barbed wire of nuclear power plants. Yes, it was an alternative to the ugly benches gouged into the hillsides of Appalachia left from the search for coal. And yes, wind was an alternative to the hodge-podge of nodding pump jacks, pipes, and oil sumps that disfigured the landscapes of more than half the states west of the Mississippi. But

Copyright © 2002 by Academic Press.
All rights of reproduction in any form reserved.

wind energy's promise was more than that of just another, although more benign, technology for exploiting nature's resources. Wind offered the prospect of a more enlightened exchange between our industrial society and the world around us: a newfound respect for the land and for the people who live on it.

This dream of wind's promise, albeit somewhat naive, envisioned an emerald city shimmering in the distance, where residents breathed clean air, drank clean water, and lived in harmony with their environment and, equally important, with each other. The wind turbines that helped power this city were clean, quiet, safe, and welcomed: symbols as well as artifacts of a choice well made.

Unfortunately, like Dorothy's Emerald City, reality has a habit of seldom living up to our expectations. When I arrived in Tehachapi that spring to see hundreds of turbines lining the pass through the mountains, I felt both excitement and disappointment. I had an uneasy feeling, a sense that something was amiss. It was to be an uneasiness that gnawed at me for the next 10 years as I struggled to reconcile the needs of business and industry with my dream.

Wind turbines were not new to me. I had seen plenty of them by the time I reached Tehachapi. I was salvaging 1930s-era wind-chargers in Montana when most of Tehachapi's entrepreneurs were still in business school. Four years before moving to Tehachapi, I had been photographing wind turbines on the first of many field trips to Denmark, the birthplace of modern wind energy. One aspect of the Tehachapi Pass that bothered me was that many of the wind turbines there—unlike those in Denmark— were not working. One of the turbines I saw in Tehachapi was simply a sham, a Potemkin turbine with a wooden board for a rotor and empty space inside its nacelle. There also was the unsettling way the earthen benches cut into the hillsides by wind developers resembled the land scarred by mining and logging elsewhere in the West; not exactly an image to win the hearts and minds of environmentalists.

The lucrative tax credits that fueled the gold-rush atmosphere at the time, plus the feverish erection of wind turbines, soon sparked a firestorm of opposition. Public meetings were crammed with standing-room-only crowds and sometimes degenerated into shouting matches between developers and residents. In Tehachapi, the local chapter of the Sierra Club entered the fray by calling for regulations to protect the environment from being plundered by this new extractive industry.[2]

This certainly was not the outcome I had foreseen, and it forced me into a decade-long search for answers to my own questions and those of others about this fledgling industry. What do people think when they see wind turbines? Why are some people disturbed and others not? What, if anything, is considered disruptive to those who live nearby? What was the experience in other countries? Were there parallels between the problems faced by wind energy and those experienced by other technologies? Were there ways to create a greater harmony between wind turbines, the public, and the landscape?

It was only after I began traveling extensively in northern Europe, especially in Denmark, Great Britain, and Germany, that I identified the themes I will be reflecting here. Most significantly, the turbines and the way they were placed in the landscape contrasted sharply with their counterparts in California. For a host of political and cultural reasons, the northern Europeans had done it differently, and they had simply done it better.[3] They offered a model different from California's "extractive" form of wind development. They showed that it was no longer necessary to turn a blind eye toward California's excesses. Wind energy could be developed with greater sensitivity and with actions, not merely words, that responded to the public banging on the door.

My preference in addressing wind energy landscapes is not philosophical or historical. I leave that to others in this volume. My intent here is to suggest pragmatic guidelines for how the wind industry and proponents of renewable energy can present wind energy's *best face*. These guidelines are culled from more than two decades of observing and photographing wind turbines and talking with scores of people about their views. The guidelines focus on visual aesthetics. The broader issue of wind energy's overall environmental impact has been discussed elsewhere.[4]

These suggestions are not meant as a guide for how publicists can deceive people, or for how promoters can conceal wind energy's defects from scrutiny, but rather as a means for presenting the dream in its most exemplary light: a means to soothe the technological edge of the soft path. Fundamental to this is the belief that renewable technologies, especially wind and solar energy, should affirm their intrinsic promise and not restate the past two centuries' paradigm of exploitation.

Although wind turbines are not necessarily intrusive, they can be. Simply stated, the objective of wind developers should be to minimize the conspicuousness of wind turbines, because people often associate conspicuousness with intrusiveness. Another objective is to lessen the "footprint" of wind energy on the land by minimizing the visual intrusiveness of access roads and other infrastructure, as well as by reducing the more familiar environmental impacts of accelerated erosion and the destruction of wildlife habitat.

Though some of the guidelines I will be proposing apply equally to individual wind turbines and to clusters of wind turbines, most apply to large arrays. The reason for this is simple: they come mostly from California's experience, and the "California model" of wind development is one of massive arrays. For example, there is one wind farm in the Tehachapi Pass that contains more than 1000 wind turbines on adjoining square-mile sections of land. Yet, and this is something I will emphasize repeatedly, there are alternative models for deployment. Wind power need not be imprisoned in large geometric arrays. In contrast, more than two-thirds of the wind development in Denmark and Germany consists of single wind turbines or small clusters. They are dispersed. Many, though certainly not all, of the environmental objections to wind energy in California would have been avoided if Americans had followed the Danish or German model.

Wind energy suffers from a high level of vague opposition. Public support often erodes once specific projects are proposed. Because support is fragile, it should not be squandered by ill-conceived projects. The wind industry and its proponents must do everything possible to ensure that wind turbines and wind power plants become good neighbors. To do so, it is necessary, as Laurie Short argues elsewhere in this book, to incorporate aesthetic guidelines into the design of wind turbines and wind power plants. It should be added that this must be done from the beginning, for once they are installed in the field, it is usually too late to correct poor design and faulty planning.

WHY DESIGN FOR AESTHETICS?

Public opinion surveys on both sides of the Atlantic Ocean have consistently shown strong support for the development of wind energy. Typically two-thirds to three-fourths of those polled—even those in areas with existing wind turbines—support wind development.[5] However, surveys in California's Altamont Pass and in the Netherlands show a tendency for those favoring wind energy to become less supportive once specific projects are proposed and wind's local impacts become more tangible. In other words, support for wind energy is strong in principle, but weakens when it is "in my backyard."

Despite its relatively benign nature, wind energy is not without local environmental and social impacts. Although these impacts may be significantly offset by the larger-scale environmental benefits of reduced emissions of global warming gases and other air pollutants, this offers little consolation to those who must live near the wind turbines. For neighbors, the impacts are immediate and obvious, the benefits distant and less visible. Neighbors absorb all the impacts, but glean only a fraction of the global gains that accrue to society at large.

In another context, that of selling so-called "green electricity" to those willing to pay a surcharge to support it, this effect has been called the "free rider syndrome." Over time it can cause resentment and ultimately a feeling of exploitation for those paying the full price for a product when they realize they are subsidizing others who share in the rewards of global environmental benefits, without carrying their share of the financial (and perhaps visual) burden.

If wind energy is to become accepted, and even welcomed, by those who live with it, the wind industry must strive to be a good neighbor. As Tip O'Neill, the former Speaker of the U.S. House of Representatives, was fond of saying, "All politics is local." And although society at large may deem wind energy desirable, if those who view or hear it nearby believe otherwise, wind energy development can be stymied. Potentially more damaging to the proliferation of wind power is the long-term erosion of general public support that frequent local conflicts entail. It would be dangerous for wind energy's advocates to forget the lessons of nuclear power: decades of vicious battles over siting nuclear plants have nearly erased whatever support once existed in countries such as the United States and Germany.

Certainly part of being a good neighbor is to evaluate carefully the intrusion that wind turbines may constitute in a community. This intrusion is often described in terms of the visual change in the landscape. But intrusion can go beyond the obvious visual impacts that wind turbines produce, encompassing a host of other human responses. With this in mind, I offer the following guidelines concerning the visual aspects of wind turbines.

WIND ENERGY AND ACCEPTANCE

Some of my guidelines are based on the ground-breaking work of Robert Thayer and his team from the University of California, Davis. Thayer defined the NIMBY syndrome as finding a technology acceptable in one's county or region, but unacceptable within 5 miles of one's home. In his surveys, Thayer found that only 9 percent thought wind plants completely unacceptable. By contrast, one-fourth found fossil-fired plants unacceptable in the county, and nearly half found nuclear plants unacceptable. But wind drew the greatest NIMBY response[6] (Figure 9.1)

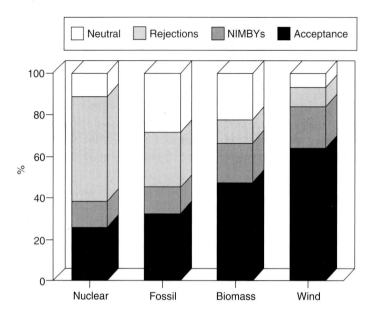

FIGURE 9.1   Power plant acceptance. (Data provided by Robert Thayer.)

The reasons behind such responses were intriguing. In the Altamont Pass, Thayer's surveys found that it is the visual intrusion or loss of visual amenities that elicits the greatest concern. Indeed, the principal impact of wind power is clearly visible for all to see, for wind turbines cannot be hidden or camouflaged. In his Solano County survey, Thayer found that the visual "quality" of wind energy garnered less support than any other aspect, even though respondents still preferred wind energy to other technologies[7] (Figure 9.2).

Maarten Wolsink has also observed the NIMBY phenomenon in the Netherlands. Wolsink found that a negative view of wind turbines on the landscape is the most significant factor for those who register opposition to wind energy. Other less significant factors included a general doubt that wind turbines would improve air quality significantly, and the fear that wind turbines would harm residents.[8] Although opposition comes primarily from seeing wind turbines on the landscape, Wolsink thinks that people unconsciously realize that opposition on aesthetic grounds is subjective and therefore often dismissed by public officials. They then rationalize their opposition by citing concerns about noise, shadow flicker, and the number of dead birds, all of which can be objectively evaluated. Despite all these other objections, visual intrusions remain the root cause of opposition.[9]

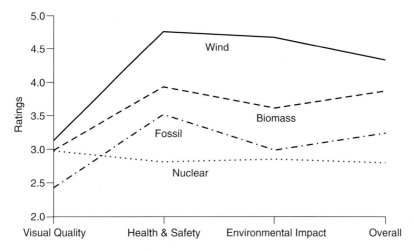

FIGURE 9.2   Power plant preference. (Data provided by Robert Thayer.)

The knowledge that a wind turbine will soon be installed in a nearby neighborhood seems to make people slightly less positive, says Wolsink. This is a near-universal response, regardless of whether respondents live in the Netherlands, Britain, or the United States. Despite the fact that in the Netherlands, 90 percent of those surveyed reacted positively to wind energy, that support is tenuous and is influenced by the distance between the respondent and the nearest turbine, according to Wolsink. The closer people live to proposed turbines, the less likely they are to endorse a proposed project. Even though 90 percent may support a project, Wolsink warns against complacency. The other 10 percent is unsupportive from the start, and it only takes one determined adversary to delay a project. Local political support is crucial, says Wolsink, but not alone sufficient for success.[10]

Public opinion shapes policy, while aesthetics shape opinion. Thus, it behooves engineers, turbine designers, project planners, and developers alike to incorporate a broad range of aesthetic factors into their deliberations. Striving to maximize acceptance is at the heart of the process of becoming a good neighbor. Maximizing acceptance is not layering a veneer of glossy public relations and hype on ill-conceived projects. It is taking public concerns seriously.

In order for a wind project to succeed, says Thayer, wind developers "must somehow enfranchise their 'visual consumers'—those neighboring residents who will be looking at the wind turbines in their landscape."[11] Thayer's comment reflects an outlook more common in Europe than in the

United States. For many Europeans, the visual resource or visual amenity belongs to the public, and its use implies an obligation to treat this public resource wisely. That is not to say that such a world view is unknown in the United States; the literature of the American environmental movement is replete with references to the use of "public resources" that are owned privately or shared in common. In the 1960s the United States pioneered laws regulating the use of public resources, such as clean air and water, for private ends. However, some leaders of the American wind industry unfortunately find such statements anathema. Like the robber barons of the 19th century, they view public resources only for the private taking.[12] It is this disregard of their social and environmental obligations that has cost the U.S. wind industry, especially in California, so dearly. Good neighbors, which wind companies have not always been, carefully consider how their private acts affect those around them. Good neighbors do not pound their fists on the boardroom table and declare "the public be damned."

There may be no way to eliminate every objection to the appearance of wind turbines on the landscape. However, there is some consensus on how to minimize these objections. These guidelines can be as simple as those of Lex Arkesteijn, who reduces the lessons he has learned from developing projects in the Netherlands to two simple commandments: build an aesthetically attractive project, and keep the turbines turning.[13] Another simple suggestion has come from the Løgstør district council in Denmark: all turbines should look alike, and they should all rotate in the same direction.[14]

WHAT WE CAN DO

If we take clues from the experience in the United States and Europe, here is what we can do to reduce the objectionable aesthetic impacts of wind power.

Provide visual order. The absence of visual order is the principal aesthetic criticism of California wind farms. They are often described in terms of the "disorder, disarray, or clutter" of turbines on the landscape. Maintaining order and visual unity among clusters of turbines is the single most important means of lessening the visual impact of large arrays. For example, landscape architecture students from California State Polytechnic University at Pomona concluded that if developers were simply to use only one kind of turbine in each project, they would substantially reduce the visual clutter evident at California sites in the early 1980s[15]

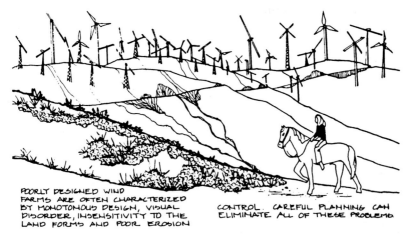

POORLY DESIGNED WIND
FARMS ARE OFTEN CHARACTERIZED
BY MONOTONOUS DESIGN, VISUAL
DISORDER, INSENSITIVITY TO THE
LAND FORMS AND POOR EROSION

CONTROL. CAREFUL PLANNING CAN
ELIMINATE ALL OF THESE PROBLEMS

FIGURE 9.3   Visual clutter. Illustration of the aesthetic problems architecture students found in a 1984 study of California wind power plants. The jumble of different types and sizes of wind turbines creates visual chaos. The students also found erosion scars from improperly built and poorly maintained roads visually disruptive. (Originally published in R. Fulton, K. Koch, and C. Moffat, "Wind Energy Study, Angeles National Forest," Graduate Studies in Landscape Architecture, California State Polytechnic University, Pomona, CA, June 1984.)

(Figure 9.3). The objective is to encourage the eye to follow across a line of wind turbines without abruptly halting at a visual interruption. This prevents the "missing tooth effect," where the observer focuses on the disruption or missing tooth, and not on the previously uninterrupted sense of order. Visual interruptions can take many forms. One example is an array where adjacent or nearby rotors spin in opposite directions. Another once-common example in California came in the form of arrays that interspersed two-bladed turbines among three-bladed machines, or turbines on tubular towers with those on truss towers (Figure 9.4).[16]

Provide distinct visual units. Studies both by American and British teams proposed the need for visually distinct groupings of wind turbines when placed in arrays. Long lines of turbines or large arrays should be separated by open undeveloped zones to create distinct visual units (Figure 9.5).[17] This also prevents the "cluttering" effect seen on hillsides in the Tehachapi Pass and once common in the San Gorgonio Pass near Palm Springs.

Provide visual uniformity. Even when large numbers of turbines are concentrated in a single array, or there are several large arrays in one locale, visual uniformity can create harmony out of a potentially disturb-

FIGURE 9.4   Visual clutter. Whitewater Wash, spring 1997. Visually disturb-
ing mixture of two- and three-bladed turbines, truss and tubular towers, working
and nonworking wind turbines. Many of these turbines, 13 years after the
problem was first identified, were finally removed in 1998. (Telephoto lens
foreshortens distance.) (Courtesy Paul Gipe.)

FIGURE 9.5   Visual units. Architecture students' view of how best to cluster
wind turbines into distinct visual units. Note absence of cut banks, fill slopes, and
erosion scars on road traversing the slope. (Originally published in R. Fulton, K.
Koch, and C. Moffat, "Wind Energy Study, Angeles National Forest," Graduate
Studies in Landscape Architecture, California State Polytechnic University,
Pomona, CA, June 1984.)

FIGURE 9.6   Linear uniformity. Micon wind turbines aligned along canal at Eemshaven, in the Netherlands' Groningen province. The uniformity of the turbines and their linear alignment along the canal make a powerful visual statement. Large natural gas-fired power plant in the background. (Courtesy Paul Gipe.)

ing vista. Visual uniformity is simply another way of saying that the rotors, nacelles, and towers of all machines in an array should appear similar (Figure 9.6). They need not be identical. There are four different types of wind turbines among the 100 machines at Tændpibe-Velling Mærsk on the west coast of Denmark (Figure 9.7). Yet all the turbines appear similar: they all have three blades, white nacelles, white tubular towers, and their rotors spin in the same direction. As a result this site is one of the world's most visually pleasing wind power plants.

Use similar turbines and towers together. One study of California's San Gorgonio Pass warned against extensive mixed arrays.[18] It recommended that if a project begins with a wind turbine that uses three blades, all subsequent turbines installed nearby should also use three blades. If the initial turbines use truss towers, all turbines added later should also use truss towers. If the nacelle has a distinctive shape, all turbines used to expand an array should use a similar nacelle. Likewise, all the turbines should spin in the same direction.

Use towers of consistent height. One should adhere to this principle unless the array is part of an aesthetic whole, such as in a wind wall of

FIGURE 9.7   Visual uniformity. This installation in Denmark remains one of the world's most visually pleasing, in part because all the turbines appear similar. (Courtesy Paul Gipe.)

wind turbines on towers of staggered height. For example, the Cal Poly study examined a proposed development in the Angeles National Forest, and then suggested that varying height can add visual interest to an array, but only if designed as a whole.[19] In other circumstances, towers of seemingly random heights destroy any uniformity that otherwise might exist. In San Gorgonio Pass, Riverside County's study complained that the sole distractions from the horizontal mass of machines on the Whitewater Wash near Palm Springs were the array of Carter turbines, which stood out on guyed towers twice the height of the others around them[20] (Figure 9.8).

Limit the number of turbines per cluster. As a means of providing distinct visual units, some groups are suggesting limiting the number of wind turbines in a cluster. Although some landscapes may be able to absorb large arrays, there is a growing consensus, especially in Europe, that small clusters are preferable. Of significance for the future of massive California-style arrays, a survey in Great Britain found that acceptance decreases with an increasing number of turbines. Projects with more than 50 turbines were acceptable to fewer than one-fifth of the people surveyed. In Cornwall, about one-third of those who did not object to wind turbines at Delabole, the site of Britain's first wind plant, said an acceptable number was "as many as possible," but a majority of the nonobjectors

FIGURE 9.8   Inconsistent tower heights. Carter wind turbines on the White-water Wash near Palm Springs stuck out "like a sore thumb" above the mass of turbines on the dry riverbed. Worse, the Carter turbines seldom operated, drawing attention to themselves. These photos were taken 10 years after the turbines had been installed. All have since been removed and sold for scrap. (Courtesy Paul Gipe.)

picked arrays of 6 to 10 turbines.[21] Anecdotal reports and the positions of environmental groups in Denmark and Germany reflect a similar preference on the Continent as well. Germany's largest environmental organization, BUND (Bund für Umwelt und Naturschutz Deutschland), wants to limit large megawatt-size turbines to clusters of 2 to 5 units, and 500 to 600-kW turbines to groups of no more than 10 turbines.[22]

Use open spacing. To avoid the dense visual clutter typical of California's wind turbine landscapes, designers should use greater spacing among the turbines. The public finds open arrays less threatening than the dense forest of turbines once seen on the floor of the San Gorgonio Pass.[23] Despite the worldwide trend toward larger wind turbines in more open arrays, dense arrays have not been abandoned, especially in the United States. In Texas, where there is much less attention to land-use planning regulations than is common elsewhere, one wind company has packed modern 700-kW turbines in an array nearly as dense as those seen in California.

Keep them spinning. When wind turbines are seen spinning, they are perceived as functioning, and therefore, beneficial. Observers are quicker to forgive the visual intrusion if the wind turbines serve a purpose; and this they can do only when they are spinning. When significant numbers of turbines do not turn when the wind is blowing, the simplest expectation of the observer is violated, says Thayer.[24] Even those opposed to wind energy often note that they would moderate their position if the turbines were seen spinning more often.

Remove nonoperating wind turbines. Reviewing comments from respondents in his Altamont survey, Thayer concluded that inoperative turbines equaled or exceeded siting, design, and scenic characteristics in causing negative responses. Thayer deduced that the single most significant action California wind companies could take to boost public acceptance was to quickly fix broken turbines and remove those that were unrepairable.[25] Yet by 1991, there were still enough derelict turbines near Palm Springs alone for the Edison Electric Institute's Charles Linderman to plead with the American Wind Energy Association, "Please get those inoperative machines down, to avoid the misinterpretation that wind still doesn't work."[26] Fortunately, by the end of 1998 nearly 1000 of the 3500 wind turbines which once stood in the San Gorgonio Pass had been removed. Some were replaced with fewer, but larger and more reliable, turbines. For example, in one project alone, the 85 troublesome wind turbines on truss towers pictured in the movie *Rain Man* were replaced with seven large, sleek turbines on tubular towers.

Use only free-wheeling rotors. Some early wind turbine designs, such as the Enertech and ESI turbines, used their generators to motor the rotor up to operating speed. In light winds these turbines could consume more energy than they produced by starting and stopping frequently. To prevent this, designers set the threshold startup wind speed higher than on comparable wind turbines whose rotors free-wheeled up to their operating speed. On the floor of the San Gorgonio Pass's Whitewater Wash, for example, it was easy to spot these early American-designed turbines in light winds because their rotors were typically not turning while the sea of European machines surrounding them was awash in spinning rotors. Even when the turbines were fully operational and not broken, they would more frequently appear idle to passers by than the free-wheeling European turbines. These manufacturers are now out of business, however, and their designs are all headed for the scrap heap. Nearly all commercial medium-sized turbines today employ free-wheeling rotors.

Remove headless horsemen. A phenomenon related to derelict turbines is that of "headless horsemen." In an effort to squeeze every last cent out of the aging stock of wind turbines in California, some operators have resorted to scavenging parts from their existing fleet. When the rotor and

FIGURE 9.9   Headless horseman. Crumpled nacelle of a WindMaster turbine lies at the base of its headless tower in The Altamont Pass during spring 1997. In the background, another WindMaster stands idle with a broken and dangling blade. (Courtesy Paul Gipe.)

nacelle are removed and valuable components scavenged for use in repairing the remaining machines, the tower is left standing "headless," without its nacelle. Some projects are dotted with these towers. Not only must operators remove inoperative turbines as soon as possible, they should remove the tower as well (Figure 9.9).

Remove ancillary structures. One of the striking contrasts between wind power plants in Britain and those in California is the general absence of buildings, power lines, and storage yards. The British architectural firm retained by Wales' Dyfed county advised that nearly all ancillary structures should be removed from hilltop sites to avoid cluttering the skyline (Figure 9.10).[27] For the most part, a visitor to a British wind farm will find only wind turbines, a farm track, and sheep. This, unfortunately, is not the case at some modern projects in North America. Enron transplanted some of its poor site practices from California to a vast project near Storm Lake, Iowa, where transformers, inverters, pendant power cables, and other electrical equipment add to the confusion produced by row upon row of lattice towers.

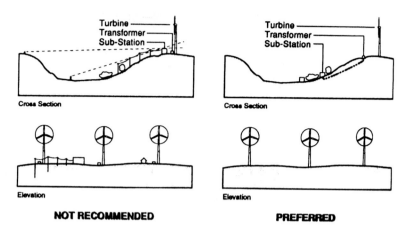

FIGURE 9.10   Ancillary structures. Removing transformers, substations, and buildings from hilltop sites reduces visual clutter, as does burying transmission lines. (Used with permission of Chris Blandford Associates.)

Bury intraproject power lines. Thayer reached conclusions similar to those of his counterparts in Britain. From his surveys in California, he recommends that developers bury all power lines and integrate extraneous equipment, such as transformers, into the turbines themselves or remove them from the site.[28] The latter is now possible with the advent of larger turbines. When the larger turbines are used with tubular towers, the transformers and control panels can be installed inside the towers, as is done on offshore and harbor breakwater installations. With the exception of some continued use of pad-mounted transformers, these measures have become common practice at British wind installations (Figure 9.11). In the United States, however, it is still not a uniform practice. It was not, for example, incorporated in Minnesota's large modern wind projects built in the late 1990s. And at Northern State Power's Phase I and Phase II projects where all intraproject power collection cables were buried, the transmission lines leaving the sites were carried overhead along rural roads. As a result, anyone viewing the projects from these public roads must look through power lines.

Harmonize ancillary structures. In Britain and Italy, wind projects go beyond Thayer's recommendation. When ancillary buildings are necessary, developers construct them of local materials to harmonize their structures with those that are an accepted part of the landscape. For example, both Renewable Energy Systems, at Carland Cross in Cornwall, and EcoGen, at New Town in Wales, used native stone for the façades and

FIGURE 9.11   British wind farm. No power lines or other ancillary structures clutter the view of the wind installation at Haverigg in Cumbria county on the west coast of northern England. (Courtesy Paul Gipe.)

slate for the roofing of their substations. These features match traditional, indigenous building styles. In another example, Italy's national utility, ENEL, used flagstones to pave access to its turbines at Acqua Spruzza in the Apennines, thereby hardening the footpath and reducing soil disturbance. ENEL also used native stone for the façades of their control buildings (Figures 9.12 through 9.14).

Avoid mounting telecom antennas. It is equally important that wind turbine owners avoid attaching telecommunication dishes, antennas, or cellular telephone repeaters to towers (Figure 9.15). Some wind turbine operators in Denmark, Germany, and the Netherlands have subcontracted space on their towers to cellular phone companies, earning themselves a few thousand dollars per year with little effort. (A rotor on one turbine in the Netherlands, however, was damaged by the crane used to mount the repeater.) Although most of these installations do not detract from the lines of turbine and tower as much as this ungainly antenna installation on an Enercon turbine in New Zealand, they certainly mar the overall appearance of the wind turbine and lend support to the charge that a wind turbine is just another industrial structure.

FIGURE 9.12 Harmonizing ancillary structures. National Windpower's transformer building at Kirkby Moor in Cumbria, northwest England. The façade uses native stone and the roof is covered with slate. Note windswept tree at left and wind turbine protruding just above the crest of the hill. (Courtesy Paul Gipe.)

FIGURE 9.13 Hardened pedestrian access. The "Appian Way" of wind energy at ENEL's Acqua Spruzza test field in Italy's central Apennines. The façades and roofs of the control buildings use native stone, and the footpaths to the turbines are paved with flag stones. The flagstones "harden" the walkway, limiting erosion from foot traffic. (Courtesy Paul Gipe.)

FIGURE 9.14 Hardened vehicle access. Hardening the access tract is also used at a small wind plant of six turbines along a canal in the Wieringermeer polder in north Holland. The prefabricated concrete blocks are designed to permit water percolation (a common feature of sidewalks and parking lots in lowland Europe) while supporting wheeled traffic. Nearby pad-mounted transformer is shrouded with a pleasing stone finish. (Courtesy Paul Gipe.)

Minimize earth moving and control erosion by avoiding steep slopes. Anytime a bulldozer operator drops his blade and plows across the landscape, he leaves a scar. And in arid parts of the world such as the western United States, such scars can remain visible for generations. Although wind energy may be relatively benign, bulldozers are not. From most vantage points, road construction brings unwanted attention to wind energy, most notably when it is in steep terrain where the cut bank and

FIGURE 9.15   Telecom antennas. Awkward installation of a telecommunications antenna on an Enercon E40 tower at Wairarapa, New Zealand's first wind power plant. The E40, with its large ring generator, is ungainly enough on its own, but the antenna platform accentuates its industrial appearance. (Courtesy Paul Gipe.)

spill slopes provide a dramatic contrast with undisturbed landscapes, often leading to accelerated erosion.

Environmentalists' distaste for such erosion includes the scars themselves, as well as the increased siltation of stream beds, alteration of stream courses, and increased flooding that can accompany it. The rill and gully erosion seen in the Tehachapi Pass has left deep cuts in the surface of the landscape. More galling than the erosion itself is the abuse of the soil resource it represents, because it is almost always unnecessary and avoidable.

The wind industry must pay more respect to the land, or it will certainly suffer further at the pen of activists such as the Audubon Society's Steve Ginsberg, to whom such erosion "is just one of many egregious examples of how wind energy is ripping up the Tehachapis," which exhibits the industry's "lack of true environmental concern"[29] (Figures 9.16 through 9.18). Ginsberg is not alone in his views. Landscape architect John Lyle, an advocate of sustainable development, has also called attention to the problem, suggesting that some San Gorgonio Pass slopes were unsuitable

FIGURE 9.16 Tehachapi Pass erosion scar. Gully cutting into steep slope on Cameron Ridge below the former FloWind site in 1995. After this photo was taken, the Darrieus turbines were removed and new turbines installed. The new owners brought in British engineers to manage redevelopment. Local activists found the British engineers more sensitive to environmental concerns than domestic wind companies. (Courtesy Paul Gipe.)

FIGURE 9.17 Gaping gully cutting across the Pacific Crest Trail on Cameron Ridge in the spring of 1997. The gully results from runoff leaving a wind farm. (Photo shows the late Keith Dawber of Dunedin, New Zealand.) (Courtesy Paul Gipe.)

for wind energy.[30] To others, such as Howard Wilshire of the U.S. Geological Survey, roads and the erosion they cause are *the* principal environmental impact of wind energy.[31]

Minimize or eliminate roads. Wind companies can reduce the risk of serious erosion by minimizing the amount of earth disturbed during construction, principally by eliminating unnecessary roads, allowing buffers of undisturbed soil near drainages, ensuring revegetation of disturbed soils, and designing erosion-control structures adequate to the task.

The single most reliable technique for limiting erosion is to avoid grading roads in the first place. Indeed, roads are not absolutely necessary.

FIGURE 9.18   Mountaintop removal? Construction on this steep ridge at Zond's Victory Garden site required blasting and bulldozing pads for erecting the wind turbines. This photo, taken in 1986, is reminiscent of the "mountaintop removal" method of mining found in the coal fields of Appalachia. The wind turbines on the right have since been removed because of poor wind conditions. As of 1998 the foundations had not yet been removed. (Courtesy Paul Gipe.)

There are, for example, no service roads to the turbines at the Tændpibe wind plant near Ringkøbing in Denmark. There are a few existing farm roads in the vicinity, but the majority of the wind turbines are serviced by special all-terrain vehicles. This may be the preferred approach in the United States as well. Glenn Harris, a biologist for the U.S. Bureau of Land Management's Ridgecrest (California) office, suggests that driving overland to install and service turbines, rather than grading roads, would significantly lessen erosion damage in arid lands.

Use existing roads. Wherever possible, developers should use existing roads or farm tracks instead of grading new roads. One of the distinctive features of British wind plants is the scarcity of roads. Planners encourage British wind companies to use existing farm tracks as much as possible. This "tread lightly" practice minimizes the scarring and erosion caused by road construction (Figure 9.19).

Minimize grading width. American wind companies typically grade roads twice as wide as those found on British, German, and Danish wind

FIGURE 9.19   Minimizing roads. Many European wind plants use existing
roads for access to the wind turbines. A rural road passes by the turbines at Royd
Moor in Southern Yorkshire. (Courtesy Paul Gipe.)

farms. American bulldozer operators typically make two passes, resulting
in roads that are twice the width of the dozer's blade. Wide roads allow
large trucks to pass unimpeded. This allows rapid, end-of-the-year, tax-
credit-driven construction to proceed. British wind developers use a
system of parking bays, or "laybys," which allow construction traffic to
move freely but with far less bulldozing and surface disturbance.

Minimize staging areas and crane pads. Staging areas are temporary
facilities for assembling towers and rotors. Crane pads are used as a
platform for the large cranes used to erect the tower and turbine. In the
United States and in some other countries, staging areas and crane pads
are bulldozed to bare earth. All vegetation is scraped clean, and the site is
leveled. Creating crane pads can lead to extensive earth-moving, and in
steep terrain these pads add significantly to the total amount of a land
surface disturbed by construction. Some staging areas in the Tehachapi
Pass and many crane pads have never been revegetated since development
began nearly 20 years ago. During heavy rainfall these areas shed runoff,
leading to erosion.

Restore original contour and revegetate. Disturbed surfaces should be
restored as closely as possible to their original contour and revegetated

immediately after construction is complete. On large projects, it is some-times possible for this to be done contemporaneously with construction. Prompt revegetation not only will limit erosion, but also will begin restoring the preconstruction color and texture of the landscape. Many of the farm tracks and roads used during construction in Britain are covered by grass within the first 1 to 2 years after the turbines are installed. Sheep graze the roads soon after construction is completed. The federal Surface Mining Act, which regulates strip mining of coal in the United States, requires miners to revegetate mined lands and restore the land to near its original contour. If the coal industry can meet such requirements, surely a more benign technology can readily do so as well.

Be unobtrusive. This is a broad category that simply suggests avoiding features such as flashing lights, signs, or painting schemes that garishly call attention to the turbines.

Avoid aircraft obstruction markings. Though the wind industry cannot make its turbines disappear, every effort should be made to avoid heightening contrast. Aircraft obstruction marking of tall structures, is, by definition, intended to increase the contrast between the structures and the landscape, so that pilots can see and avoid them. To remove the association of wind turbines with other industrial structures requiring obstruction marking, such as smokestacks and telecommunications towers, designers must limit the height of wind turbines and should avoid sites near airports where aviation regulations require obstruction marking either with alternating red and white bands or with flashing lights (Figure 9.20).

Douse security lights. Operators should douse security lighting at their wind plants and substations, in order to decrease the contrast between the wind plant and the nighttime landscape of rural areas where wind turbines are typically installed. Lights disturb the tranquility of the night sky. Nighttime security lights are nonessential and can be activated as needed by motion detectors such as those used by Southern California Edison in light-sensitive residential neighborhoods.

Avoid billboards. All signs near a wind turbine or at a wind plant should serve solely to inform the public about the wind turbines and their place on the landscape. Operators should avoid using wind turbines as a means for elevating advertising billboards to new heights. Billboards, like any other extraneous structure, detract from the impression of purity that wind turbines should ideally impart to the viewer. Billboards add visual clutter to the landscape.

Avoid logos on nacelles. In much the same way, wind turbines need not advertise for their manufacturer or for their sponsors across the country-

FIGURE 9.20   Obstruction markings. Painted with alternating bands of red and white. Hawaii. (Courtesy Paul Gipe.)

side. Society accepts, sometimes begrudgingly, wind turbines' visual intrusion into the landscape for the purpose of producing clean, renewable, wind-generated electricity, not the promotion of the wind developer or wind turbine producers. The public is less willing to accept, and can even be offended by, a company's advertising logo emblazoned on the side of 100 nacelles, each the size of a large truck. For this reason, some planning authorities prohibit logos on nacelles, though they may permit nacelles to bear more discreet identification legible from the base of the tower.[32]

Choose color carefully. Wind turbines will always be visible on the landscape. This cannot be avoided. No amount of camouflage will make wind turbines invisible. But color can be important in reducing impact. A light tan often works best in arid environments, while a light gray or off-white can be the best choice in temperate climates. However, there is disagreement about what is the most acceptable color. One school of thought argues that the color of wind turbines should not contrast sharply with the surrounding landscape. This leads to the use of gray or off-white paint. Others argue that since wind turbines cannot be hidden, there should be no attempt to obscure them. Following such reasoning, using a low-contrast color scheme is a subtle attempt at camouflage, or worse, a form of public deception. Since the turbines cannot be hidden, the argument continues, they should boldly acknowledge their presence with white towers and nacelles. Vestas, a Danish wind turbine manufacturer, uses white. Bonus, another Danish wind turbine manufacturer, uses gray or off-white towers and nacelles.

In Denmark, white is an accepted color on the rural landscape. White stucco is found on pre-19th-century buildings, especially on old *gårds* (farms). The ubiquitous Danish flagpole is also white, not gray.[33] Despite Danish planning authorities' acceptance of both white and gray, they prohibit the use of other colors, notably blue.[34]

Although white may present a higher contrast than gray on northern European landscapes, it is a symbol of purity that conveys an intrinsic and powerful message about wind energy. There are few views of wind energy more dramatic and yet seemingly more in harmony with the landscape than the hundreds of white Vestas turbines scattered randomly across the green fields of Syd Thy in northwestern Jutland.

Use proper proportions. Wind turbine designers should consider the appearance of their work on the landscape as part of their design criteria, alongside cost effectiveness and productivity. Rotor, nacelle, and tower should form part of an aesthetic whole. Wind turbine designers and wind power developers alike should avoid considering the wind turbine and its various tower options a mix-and-match set. Turbines and towers should form an aesthetic unit and should be designed with particular tower sizes and shapes in mind.

Some of the most pleasing wind turbine designs include the clean lines of the Bonus' Combi and the award-winning Danwin 23-meter turbine (Figure 9.21). The simple nacelle on the Folke Center for Renewable Energy's 400-kW turbine is similarly appealing. The Vestas V27 on a 30-meter (100-foot) tower is particularly attractive, and probably represents the ideal of what a wind turbine should look like. The slender, but not overly thin blades of the rotor, the clean lines of the nacelle, and the height, thickness, and taper of the tower all appear in harmony.

Not all Danish designs are so successful. Regrettably, Vestas' designers have lost their way since the V27 with the introduction of slab-sided larger

FIGURE 9.21   Pleasing proportions. Clean lines and good balance between nacelle and tower are hallmarks of the DanWin 160 kW. Kern County, California. (Courtesy Paul Gipe.)

machines. The nacelle on the V40 series is boxier, and the tower is stockier than those of the V27 series. The V65 reverses the proportions again, with a boxlike nacelle balanced precariously on the slender neck of a tall tower. Even the V27 can be misapplied when used with an exceptionally tall tower. On Alta Mesa near Palm Springs, several rows of V27s are mounted atop 50-meter towers. The balance among rotor, nacelle, and tower is lost. The towers appear too slender, almost sticklike. The taper is too strong near the nacelle, which is necessary to allow sufficient clearance between the blade tips and the tower. This effect was exacerbated when the V47 model was used with 60-meter (180-foot) tall towers in wind plants installed in the late 1990s. One cannot add a substantially larger rotor to a nacelle of fixed dimensions without upsetting the aesthetic balance among the rotor, nacelle, and tower.

That some designers, as well as customers, are sensitive to the value of appearance is showing up in the actions being taken to make their products more aesthetically acceptable. When Enron bought Germany's Tacke, for example, they immediately replaced the slab-sided angular nacelle on the 600-kW model with a more flowing form. Similarly, Enercon shrouded the ring generator on their E66 with a smooth fiberglass nacelle, obscuring the large-diameter ring generator that was once the ungainly signature of the smaller E40. Enercon has since upgraded the E40 model to include the generator shroud.

Maintain good housekeeping. A long list of items that can be used to reduce the visual clutter and disorder typical of California wind plants falls under the rubric of general housekeeping. Some things that can be done, such as adjusting visual density or choosing three-bladed turbines, are opportunities unique to wind energy. Most items on the list, however, are not. They are the prosaic prescriptions that our parents teach us as children. We learn to pick up after ourselves and to consider the effects our actions have on others. For managers of wind plants, this translates into a respect for the environment and the community of which they are a part.

Always "dress" your wind turbine properly. Wind turbines should never "go out in public" without proper attire. All wind turbines should include a streamlined nacelle cover to soften the lines between the rotor, nacelle, and tower. A wind turbine without a nacelle cover is like a car missing its hood, or a businessman without his suit and tie. The viewer quickly senses that something is amiss and is most likely to react unfavorably. Operating a wind turbine without a proper nacelle cover and nose cone (spinner) is akin to driving a car without its sheet-metal skin.

Clean nacelles and towers. Some wind turbines, such as the Kenetechs and Mitsubishis, are "incontinent," regularly spilling their internal fluids on their blades and towers. Dust and grime stick to the oil, dirtying the turbine and tower. When left unattended, the soiled turbines begin to look like props for a Hollywood movie about a post-Armageddon world. Setting an admirable example, the operators of the Mitsubishi turbines near Mojave in the Tehachapi Pass wash them regularly to remove accumulated oil, at a cost of several thousand dollars per turbine.

Such maintenance should be part of doing business. Responsible managers and wind turbine designers alike must ensure that nacelles hold all oil or fluids which are likely to leak. If they do leak, these managers should promptly clean the turbine and tower, returning the site to its pristine condition. Operators know that no manager at a nuclear power plant or an auto assembly line would long keep his job if he permitted oil to pool on the shop floor. A wind plant is no different. Operators and employees alike understand that the public intuitively judges management by how it executes simple housekeeping chores. A company's lack of concern for the obvious can indicate a disinterest in the less visible tasks, such as the safe disposal of hazardous wastes.

Keep sites tidy. All three of California's principal wind sites are semiurban. Even the Tehachapi Pass, which is 3 hours by car from Los Angeles, suffers the ills common to the urban fringe. Scattered around wind plants in all three locations are discarded beer cans, broken wind turbine blades, bits and pieces of wind turbines, rags, and other assorted detritus. On the Foras site atop Cameron Ridge in the Tehachapi Pass, pieces of fiberglass blades can be seen lodged in Joshua trees. Although the litter festooning some wind power plants in California may be part of a nationwide trend toward the "trashing of America," this offers little justification to operators for not policing their sites and removing the trash of a careless industrial society (Figure 9.22).

FIGURE 9.22   Litter. Discarded mattress along access road to the wind turbines on Painted Hills in the San Gorgonio Pass, north of Palm Springs, California. (Courtesy Paul Gipe.)

Fastidious site managers care enough to require technicians to pack their litter out with them at the end of the day, instead of allowing it to blow across the landscape. And they pick up the debris that others dump on their sites and dispose of it properly. Good managers also care enough to ensure that the turbines, where numbered, are identified with a crisp, legible stencil rather than a slovenly spray-painted scrawl. And they are never too busy watching the bottom line to notice the day-to-day details that govern how the public views them and wind energy.

Remove all bone yards. Some wind plants in the Tehachapi Pass, such as Enron's Victory Garden development, have unsightly scrap heaps or what the locals call "bone" yards. These yards contain a bewildering array of junk. Enron's bone yard is a veritable museum of abandoned wind turbine hardware. At another location, Cameron Ridge, Cannon's bone yard at one time included abandoned cars, pickup-truck camper shells, wind turbine wreckage, leaking gear boxes, scrap wood, and broken pallets (Figures 9.23 and 9.24). Fortunately, this is no longer the case on Cameron Ridge.

Bone yards do wind power no good. The public judges wind power plants in their entirety, not just on the turbines themselves. If operators

FIGURE 9.23   Bone yard I. Improper disposal at a bone yard on Enron's (formerly Zond Systems) Victory Garden site, spring 1998. Corporate office is in the background. (The large factory is a cement plant.) (Courtesy Paul Gipe.)

FIGURE 9.24   Bone yard II. Improper disposal of scrap wind turbine blades on Enron's (formerly Zond Systems) Victory Garden site in the Tehachapi Pass. There was at least one other large bone yard on the Enron site in spring 1998. (Courtesy Paul Gipe.)

allow an accumulation of wind turbine blades, nacelles, cable spools, disused tools, and other machinery, it sends a signal to the public that the operator is careless with a public resource: the visual amenity. The public expects wind energy to be a clean source of energy. If there are abandoned cars or broken wind turbines littering the site, this expectation is violated and the public becomes less sympathetic to the wind industry's use of the visual resource. They are also less likely to accept wind energy as a "green" resource.

Respect the land and the landscape. In the "disposable" society we seem to have developed, land and landscapes are often viewed as disposable as well. This can be seen in the hard rock mining landscapes of the western United States, as well as in coal mining regions of Appalachia and the Illinois Basin, where abused lands invite further abuse. Prior to the 1977 Surface Mining Act, mined lands were seldom reclaimed. These landscapes were littered with abandoned pits and high walls, broken rock, and derelict mining equipment. Since these were literally "junk" or "trash" landscapes to the mining companies, the neighboring communities viewed them similarly. These mined lands became unofficial, and sometimes official, dumping grounds for whatever society wished to discard: garbage, abandoned cars, and so on. Some poorly designed or abandoned wind plants in California have fallen into the same pattern of misuse. With road scars, broken parts, oil barrels, and derelict wind turbines, these wind plants invite the dumping of urban waste. It's common to find the hulks of abandoned cars littering some California wind farms. As if in a scene out of impoverished Appalachia, the Sierra Club, when visiting Cannon Energy's site on Cameron Ridge in the Tehachapi Pass in 1997 found—along with uncontrolled erosion and oil leaking down the tower of a turbine—a pickup truck dumped into a gully. Not far away, at an abandoned wind farm, two cars sat on their rims, their windows shot out and parts strewn about.

Although urban trash might sometimes be scattered at the gates of a national park, it is unlikely, and certainly unacceptable. Wind sites should be no different. Where there is a perception that the land and the landscape are not valued or respected, there is less reluctance to contribute to its further decline. The lesson is that if wind developers and wind plant operators respect the land, others are more likely to respect their use of it as well.

Inform the public or provide public access. Wind turbines are not inherently dangerous, and every aspect of a wind plant should convey the sense that wind energy is more benign than other forms of energy. Wind turbines and wind plants should be welcoming. Designers can accomplish this by eliminating fences and warning signs, and by providing points of public access, footpaths among the turbines, and informational kiosks. By using a public resource, the landscape's visual amenity, wind developers bear an obligation to inform the public about how they are using this public resource responsibly. The wind industry can do so by providing access and by building visitors' centers. These need not be elaborate; they can take the form of simple kiosks or even simple signs that provide basic information about the wind plant: how it works, and the contribution it makes. Many sites in Europe provide just this sort of information as well as public access (Figure 9.25).

Limit tower height and turbine size. According to Lewis Mumford's "technological imperative," if a technology exists, it will be used.[35] The classic example of this imperative is nuclear weapons. Once the United States developed nuclear weapons, we were compelled to use to them in quelling Japan. When applied to wind energy, this imperative is seen in the increasing height of towers. Taller towers increase revenue per turbine. As technological improvements make taller towers possible, developers begin using them. In the early 1990s towers typically reached heights of 30 to 40

FIGURE 9.25   Public access. Providing access to a curious public need not be elaborate. Access can be as simple as a parking area and kiosk, as at Royd Moor in central England. Many wind plants in Great Britain provide gates for sightseers and hikers. (Courtesy Paul Gipe.)

meters (100 to 130 feet); by the mid-1990s they had reached 40 to 50 meters (130 to 160 feet); and by the end of the decade towers 60 to 70 meters (200 to 230 feet) tall were common in Germany. Some towers are now football-field lengths of 100 meters (330 feet).

The use of increasingly taller towers may be one reason why wind turbines have become visible from commercial aircraft flying over Germany.[36] Tall towers permit the turbines to stand well above surrounding obstructions—trees and buildings—and the terrain. This increases their visibility. Because tall towers are navigation hazards for aircraft, aviation authorities require obstruction marking. As intended, the flashing lights or garish white and red banding increases the visibility of the turbine; it also increases its intrusiveness. Exceptionally tall towers may also be out of scale with the terrain.

German environmentalists, such as Georg Löser of BUND Baden-Wurttemberg, are now accustomed to 600–700-kW turbines and finds them in "optical balance with themselves and the landscape." He says he has "made peace with them," as long as there are "not too many at one location."[37] However, megawatt turbines on tall towers are out of scale with the landscape, says Löser, by approaching if not exceeding topographic relief of 100 meters (300 feet). Certainly they are out of scale with trees and buildings, which are only 20 to 30 meters (60 to 100 feet) in height. For this reason, Löser sees a maximum total height of 100 meters for wind turbines on land: that is, a maximum tower height of 70 meters. At particularly sensitive sites, the turbines should possibly be 10 to 30 meters shorter, he says.[38]

Extremely tall towers are, like most other aesthetic factors, not a technological necessity. Tall towers result from economic imperatives. Where community standards discourage or prohibit extremely tall towers, wind development can still proceed. The Danish manufacturer Bonus, for example, installed wind turbines on towers of suboptimal heights for Britain's National Windpower. The turbines were being installed on highly cherished uplands, and to obtain planning approval, both companies were willing to use towers shorter than the norm.

Avoid tower pedestals. Miniature ziggurats began appearing on the lowlands of Denmark and Northern Germany in the late 1990s. In the heated competitive market of Northern Europe, every meter of tower height counts, and operators strive to use the tallest tower permissible. As operators seek to maximize revenues to the fullest, wind turbine manufacturers and their dealers have begun offering "tower extenders" in lieu of adding another tower section. These tower extenders most often take the form of 1- to 2-meter-high mounds or pedestals on which the tower is erected. Some also include a concrete extension of the foundation that may more properly be called a "foundation extender."

In Denmark these pedestals are appearing for another reason: aesthetics. Planners in some Danish counties (*Amts*) require that the hubs of all turbines in arrays be at the same height. Planners in these counties mistakenly believe that a line of nacelles in a row, or in several rows, is more attractive when the nacelles are all of the same exact height rather than following subtle changes in the terrain. Whereas in Tændpibe-Velling Mærsk, the towers of the 100 turbines plunge cleanly into the ground despite slight differences in terrain and in tower height, numbers of wind turbines in the nearby wind plant at Stauning are mounted on earthen pedestals.

At Carland Cross in Great Britain, as at Kaiser-Wilhelm-Koog in Germany and Tændpibe-Velling Mærsk in Denmark, there are no surface expressions of the foundations. The tubular towers plunge directly into the ground, without the large concrete pads often seen at the base of such towers in the United States. The developer at Carland Cross, Renewable Energy Systems, achieves this effect by burying the concrete pads for all

15 machines 1–2 meters (3–6 feet) below the surface, thus allowing the farmer to plow up to the base of the towers.

Angular concrete pads exposed at the surface and the angular pedestals seen in Northern Europe break up the line of the terrain, add to the visual clutter at ground level, and prevent tillage to the base of the tower. Pedestals increase the footprint of wind energy on the landscape and interrupt the strong connection between the turbine, its tower, and the earth. Rather than the tower springing from the earth as an almost organic form, pedestals and visible concrete foundations give the installation a

FIGURE 9.26   Springing forth organically. When there is no surface expression of the foundation, as at this wind plant on the east side of the Limfjord in northwest Jutland, the wind turbines appear to spring from the earth. This creates a sensation of harmony between the wind turbine and the landscape of which it is a part. (Courtesy Paul Gipe.)

clearly artificial or industrial appearance. Wind turbines should be installed with little or no surface expression of their foundations. Towers should plunge directly into the earth (Figure 9.26).

Consider the aesthetics of small wind turbines. Because of their size, small wind turbines present far less of a visual intrusion on the landscape than do medium-size turbines. But manufacturers of small wind turbines seem even less conscious of aesthetic design than their colleagues who build larger turbines. Some models, such as the towers produced by Jacobs Energy Systems in the early 1980s, are reminiscent of childhood Erector (Meccano) sets. Most small wind turbine manufacturers could use the help of a good industrial designer.

TO SUMMARIZE: BE A GOOD NEIGHBOR

Wind energy is a rapidly maturing industry and has long since outgrown its sometimes stormy adolescence. Wind energy is now a multibillion dollar industry, and it is no longer reasonable for its advocates to excuse its youthful indiscretions. With six major manufacturers and an equal number of minor manufacturers worldwide, wind energy has come

FIGURE 9.27   Good neighbors. Wind turbines at St. Breock Down above Camelford in Cornwall, southwest England, have become not just a part of the landscape, but also a part of the community. (Courtesy Paul Gipe.)

of age and should be called to account for projects that do not fulfill the high standards expected of it. The wind industry must assume the responsibilities of adulthood, like the technologies with which it must compete.

In general, the prescriptions for optimizing aesthetic acceptance can be summarized by noting that designers, developers, and operators should be good neighbors. Only when the wind industry places as much importance on being a good neighbor as it does on aerodynamic or economic efficiency will the public welcome wind turbines into their backyards (Figure 9.27).

NOTES AND REFERENCES

1. Amory Lovins, *Soft Energy Paths: Toward a Durable Peace* (New York: Harper & Row, 1979).
2. This was particularly frustrating for me since I had been a lobbyist for the Pennsylvania Chapter of the Sierra Club prior to arriving in California.
3. California's early experience presented Europeans with an obvious lesson in how not to develop wind energy. At the 1987 European Wind Energy Conference in Leeuwarden, the Netherlands, Birger Madsen of BTM Consult flashed photos of the San Gorgonio Pass on the screen during his presentation. Madsen told the audience that "never again" should wind farms proceed as they had in California. No country in northern Europe would permit such haphazard development, he said, and that if they did, there would be such a backlash that it would doom the industry. To demonstrate how a wind power plant could be artfully built, he advanced images of Tændpibe, the forerunner of Denmark's largest array of wind turbines near Rinkøbing on the west coast of the Jutland peninsula. It was as if the Danes had set out to build a model wind power plant solely to counter the negative images from California.
4. Paul Gipe, *Wind Energy Comes of Age* (New York: John Wiley & Sons, 1995).
5. For attitudes in North America see Phyllis Bosley, "California Wind Energy Development: Environmental Support and Opposition," *Energy & Environment* 1 : 2 (1990): 171–182; and Phyllis Bosley, "A Study of Energy Resources and Issues: Perceptions and Attitudes Held by National Environmental Thought Leaders," Towson State University, Towson, Maryland, 1989; Kathleen Smith and David Loveland, "U.S. Energy Policy: The 1990's and Beyond," The League of Women Voters Education Fund, Washington, DC, 1989; M. Pasqualetti and E. Butler, "Public Reaction to Wind Development in California," *International Journal of Ambient Energy*, 8 : 2 (August 1987); 83–90. For attitudes in Europe see Brian Young, "Attitudes towards Wind Power: A Survey of Opinion in Cornwall and Devon," Energy Technology Support Unit, Department of Trade and Industry, Harwell, 1993, 40; C. Westra and L. Arkesteijn, "Physical Planning, Incentives, and Constraints in Denmark, Germany, and the Netherlands," paper presented at "The Potential of Wind Farms," European Wind Energy Association special topic conference, Herning, Denmark, 8–11 September, 1992; Kristina Freris, "Love Them or Loathe Them? Public Opinion and Wind Farms,"

paper presented at 20th annual British Wind Energy Association conference, 2–4 September, 1998, Cardiff, Wales.

6. Robert Thayer and Heather Hansen, "Consumer Attitude and Choice in Local Energy Development," Department of Environmental Design, University of California—Davis, May 1989, 17–19.

7. Thayer and Hansen, "Consumer Attitude and Choice," 20.

8. Maarten Wolsink, The Siting Problem: Wind Power as a Social Dilemma. Department of Environmental Science, University of Amsterdam, The Netherlands, undated.

9. Maarten Wolsink. "Attitudes and Expectancies about Wind Turbines and Wind Farms," *Wind Engineering*, 13 : 4 (1989); 196–206.

10. Wolsink, "The Siting Problem."

11. Robert Thayer and Heather Hansen, "Wind on the Land," *Landscape Architecture* (March 1988); 68–73.

12. This comment is gleaned from 15 years of working with the leaders of the U.S. wind industry, including private conversations and interviews, and 2 years on the board of directors of the American Wind Energy Association.

13. C. Westra and L. Arkesteijn, "Physical Planning, Incentives, and Constraints in Denmark, Germany, and the Netherlands," paper presented at "The Potential of Wind Farms," European Wind Energy Association special topic conference, Herning, 8–11 September, 1992.

14. Bridgett Gubbins, "Living with Windfarms in Denmark and The Netherlands," North Energy Associates, Northumberland, England (September 1992): 7. Løgstør is on the south shore of the Limfjord in northern Jutland.

15. R. Fulton, K. Koch, C. Moffat, "Wind Energy Study, Angeles National Forest," Graduate Studies in Landscape Architecture, California State Polytechnic University, Pomona, CA (June, 1984): 64.

16. No medium-sized, two-bladed turbines were commercially available in early 2001. Though some manufacturers continue to experiment with two-bladed turbines, it is unlikely that they will enjoy commercial success, at least on land. Lattice towers are still occasionally used. Enron installed a large wind plant using lattice towers in Iowa during the late 1990s. The bordering state of Minnesota insisted that Enron use a tubular tower for a large project installed at the same time. Minnesota specifically excluded the use of a lattice or truss tower on its wind farms.

17. See R. Fulton, K. Koch, C. Moffat, "Wind Energy Study, Angeles National Forest"; and "Landscape Impact Assessment for Wind Turbine Development in Dyfed," Chris Blandford Assoc., Cardiff, Wales, February, 1992.

18. WIMP. Phase III. Wind Implementation Monitoring Program. See the Visual Element, Draft Report, Riverside County, Riverside, CA (October 1987): C-4, 12–15.

19. R. Fulton, K. Koch, and C. Moffat, "Wind Energy Study, Angeles National Forest," 64.

20. WIMP, Phase III, "Wind Implementation Monitoring Program," Draft Report, Riverside County, Riverside, CA (October, 1987): C-4. The wind turbines referred to in this report and 300 others surrounding them have since been removed.

21. Brian Young, "Attitudes towards Wind Power: A survey of Opinion in Cornwall and Devon," Energy Technology Support Unit, Department of Trade and Industry, Harwell, 1993, 37.

22. "Fur Einen Natur- und Umweltvertraglichen Ausbau der Windenergienutzung" by Marcus Bollmann, Georg Löser, Gunter Ratzbor, Wissenschaftlichen Beirat, Bund fur Umwelt und Naturschutz Deutschland, March 28, 1998. This was a draft proposal of

BUND's position on wind energy. BUND has 300,000 members nationwide. For comparison, the Sierra Club, the largest environmental group in the United States, has 700,000 members.

23. Alexi Clarke, "Windfarm Location and Environmental Impact," Natta, Open University, England, June 1988, 64. Clarke called the Whitewater Wash a "thicket" of wind turbines.

24. Thayer and Hansen, "Wind on the Land," 68–73.

25. Thayer and Hansen, "Wind on the Land," 68–73.

26. Charles Linderman, Edison Electric Institute, statement made in his oral address to the American Wind Energy Association's Windpower 91 conference in Palm Springs, California, 24–27 September, 1991.

27. "Landscape Impact Assessment for Wind Turbine Development in Dyfed," 10.

28. Thayer and Hansen, "Wind on the Land," 68–73.

29. [29] Steve Ginsberg, "The Wind Power Panacea: Is There Snake Oil in Paradise?," *Audubon Imprint*, Santa Monica (CA) Bay Audubon, 17:1, 1–5.

30. John Tillman Lyle, *Regenerative Design for Sustainable Development* (New York: John Wiley & Sons, 1994): 69–70.

31. Howard Wilshire and Douglas Prose, "Wind Energy Development in California, USA," *Environmental Management*, 10:6 (1986).

32. Viborg Amt, Denmark.

33. Flags and flagpoles are a common sight on the Danish landscape. After the loss of Schlesvig to Prussia in 1860, there was a revival of Danish national pride in the late 19th century, celebrating Danish language and culture. This is manifest in frequent displays of the *dannebrog* or Danish flag at social occasions, such as birthdays.

34. Mie Mølbak, the planner for Viborg Amt, in an October 16, 1997 interview, disparagingly called a WindMatic on a blue tubular tower near Thisted in northwest Jutland "Madame blue" after the traditional blue Danish coffee pot. The use of the blue tower prompted the Amt or county to regulate color. The Amt would "never" permit red either, says Mølbak. She personally preferred "earth tones." Though white and gray are permissible, she noted that white and gray towers are not permitted next to one another in a cluster.

35. See, for example, Lewis Mumford. *The Myth of the Machine: The Pentagon of Power* (New York: Harcourt Brace Jovanovich, 1964) and earlier works.

36. We could clearly identify operating wind turbines on a commercial flight over Germany from Copenhagen to Barcelona. The flash of sunlight glinting off their moving blades ("disco" effect) first drew our attention. Then we gradually discerned one, two, and finally several small clusters of large wind turbines in what we took for the Eifel Mountains.

37. Telephone interview, May 1998.

38. Franz Alt, Jurgen Claus, and Herman Scheer, editors. *Windiger Protest: Konflikte um das Zukunftspotential der Windkraft* (Bochum, Germany: Ponte Press, 1998). See the chapter by Georg Löser, "Windenergie: Umweltschutz kontra Naturschutz" ("Wind Energy: Environmental Protection versus Nature Protection"), pp. 75–92.

PART

V

AFTERWORD

10

A VIEW FROM LAKE COMO

GORDON G. BRITTAN, JR.

During the discussions about wind energy landscapes at Lake Como that developed into this collection, we focused on certain themes and reached tentative conclusions—this, in spite of the diversity of the participants with regard to nationality, profession, contexts, and histories. It was a spirited 10-day gathering consisting of a rich exchange of positions and views, perspectives and ideas about the future acceptance of wind power. In this Afterword, I wish to emphasize some of the issues discussed and, in a tentative way, some of the conclusions which we drew. My goal here is to indicate them as an aid to formulating clearer policy guidelines applicable to future wind energy development. Indeed, the establishment of clear guidelines is essential to the whole planning process; without guidelines tailored to specific sites in question, any proposed project is encumbered substantially from the outset.

The group shared many beliefs about wind power. For example, there was no serious disagreement that greater use of wind energy will reduce some of the problems associated with nuclear and fossil fuels. The entire group is convinced that wind energy has a place in the spectrum of energy choices, and an important one. There was also no doubt that wind power technology is now reliable and cost-effective. And we all believed that the future task is how to legitimately counter the resistance to wind energy that has been recorded in many countries. There was not, however, unanimity as to how this was to be done.

There were essentially three areas of disagreement. The first has to do with *perspectives*. Some participants believed that the main problem the industry faces is one of mitigating the visual impacts of wind energy on the landscape, largely by way of more sensitive siting of the turbines and more involvement of the public in this process. Others, however, believed that only fundamentally reorientating the way we think about wind energy,

Wind Power in View:
Energy Landscapes in a Crowded World

Copyright © 2002 by Academic Press.
All rights of reproduction in any form reserved.

and about the turbines designed to capture it, will disarm the rather widespread opposition to them.

The second main difference among the group was in *aesthetic presuppositions*. Some took it as obvious that aesthetic judgments are in some deep sense subjective, that beauty is truly "in the eye of the beholder." Others held that in the same deep sense aesthetic judgments are objective, and that there are standards which can be the basis for decisions.

The third theme that differentiated us was the relative significance of the various *problems* to be faced in seeking to win greater public acceptance of wind energy. Four main themes were mentioned: (1) the character of the technology, (2) its deployment in the landscape, (3) the system of its ownership and control, and (4) the attitudes of people to its increasing presence.

There were variations on each of these themes. Geographer Martin Pasqualetti is a *cultural* subjectivist who believes that there is socially conditioned agreement within but not between cultures regarding what is beautiful. Engineer Martin Hoppe-Kilpper is an *individual* subjectivist who maintains that "it all comes down to a matter of taste." There were also degrees to which each person pushed his or her position. Artist Laurie Short is a *radical* individual subjectivist, convinced that it is, in fact, pernicious to seek, still more to enforce, a consensus on aesthetic questions. Landscape architect Christoph Schwahn is a *moderate* individual subjectivist, who recognizes differences in taste, but suggests they are nonetheless well worth discussing.

This said, we can locate the participants in what is admittedly an overly tidy taxonomy. Most of the discussion centered on defining the problem. What exactly is the resistance to wind power, and how is it best to be resolved? There were several different emphases. Historian Robert Righter and I, a philosopher, wanted to open up the technology option, questioning whether the Danish three-bladed turbine was the culmination of technological and aesthetic advance. Righter called for a "greater sensitivity toward the possibilities" of nonconventional wind turbines, illustrating the richness of past efforts in this direction. I urged consideration of my own small-scale soft-sail design. Paul Gipe and Frode Birk Nielsen thought that such a discussion was rather pointless, convinced that the future held no radical departure in design.

Landscape architects Nielsen and Schwahn opened up the placement issue, using such techniques as computer visualization to show how varying numbers of turbines and different kinds of arrays and placements can be assessed with respect to particular sites. Paul Gipe also initiated discussion of the deployment option, focusing more on the visual appearance of the turbines themselves (their height, their color) than on their siting.

Hoppe-Kilpper and Nielsen addressed what was labeled the *system* option, among other things advocating an equity interest in the turbines by those who own the land on which they sit. For the most part these are farmers who would then count wind energy as an additional cash crop. Such local ownership and control contrast with the widespread present arrangement in which a farmer's land is merely leased by a distant corporation. There was a strong consensus that a significant equity interest by locals can markedly influence acceptance.

Our discussion often returned to the *people* problem. Short advocated "technological fatalism," that is, the idea that we have to accept that "we cannot change the dimensions of ugliness and beauty to the point where it will affect decision-making in the placement of wind farms." Implicitly he suggested that we cannot affect the technology. If we cannot change the technology and we have few choices on deployment and ownership options, then, to quote Short, "we can only change people's aesthetic perceptions."

The *people* problem has several variations. Some thought it was a *process* problem. The point is to involve as many of those directly affected by wind turbines in the process of drawing up the rules for their placement and use. Those who emphasized process included geographer Karin Hammerlund, who urged the use of sophisticated polling techniques to sample public attitudes, and Short, whose two favorite words were "consultation" and "cooperation."

Others thought it was a matter of educating the public about the desirability of wind energy. Thus Pasqualetti's view on the vast San Gorgonio wind development is that the most important element, indeed the real power of our landscape perception, is its function in educating the public about the trade-offs, relative costs, and benefits of wind energy and competing methods of generating electricity. Hammerlund added that people will more readily accept wind energy if they know that it is merely transitional in character and will phase out as other energy alternatives possibly come online. Wind turbines, after all, are easily removed from the landscape. Still others held that wind power has a *perceptual* problem. The public must learn to consider wind turbines as ingredients in aesthetically pleasing landscapes. This will be an evolutionary process, one that will happen only as the public becomes more knowledgeable about and accustomed to their presence. Elements of this position can be found in the papers of Righter, Hammerlund, Pasqualetti, and Short.

Finally, Christoph Schwahn emphasized that the main problem has to do with efficient energy use. Strictly speaking, this is an issue with all forms of energy and not just with wind. For Schwahn, when people see the downside of *all* forms of energy production, including the visual impact of wind, they will want to conserve all the more. He wants people to see the visual *price* they must pay for profligate consumption, a view Pasqualetti stressed. In Schwahn's view, we are better off in the long run not making wind turbines and their deployment too beautiful or too remote, for in that case the necessary, inevitable steps toward greater energy efficiency will only be delayed.

It is fair to say that the technology question was not widely debated. The majority of workshop participants assumed that the Danish-style three-bladed turbine would continue to be the industry standard. Although there was some discussion of scale and its bearing on the aesthetic quality of landscapes, it was generally accepted that the trend toward larger and taller machines would continue as well. Where our deliberations centered on the turbines themselves, we all emphatically agreed that the turbines must *turn*; nothing is more destructive to public confidence than a field of broken or nonoperating machines. We also discussed color and the way in which the nacelle[1] was packaged. Put another way, most of the discussion centered on new ways of deploying and owning and controlling the turbines and on the processes by which the public's attitudes and perceptions might be changed.

The historians and humanists among us knew well enough that the consensus of a moment does not often last, and that suggestions not considered in full when they are first made often come to dominate public policy.

There are many new ideas in this collection of papers, but three ideas gained more or less general support in our discussions.

1. Placement. Wind turbine placement must always be sensitive to site. This is a common principle among architects so far as buildings are concerned. But it has yet to be universally adopted in connection with wind turbines. In the United States, in particular, landscape architects play a minimal role, and a one-size-fits-all approach is still the norm. Often the only site characteristics typically considered are strength of wind and availability of power transmission lines. Aesthetic acceptability must not be an afterthought; the success of wind energy rests at least in part on the degree to which wind turbines blend into their surrounding landscape context. Naturally, *blending in* within this context includes various support structures, the buildings in which the transformers are located, the roads connecting the turbines, the transmission lines, the way they are maintained, and the diligence with which the developers keep derelict equipment from accumulating.

The majority of us also recommended that wind turbines not necessarily be hidden or camouflaged. Indeed, there was majority consensus that the visual character of wind turbines should not be disguised. Rather, some saw their blatant display as a kind of honesty, which is an important element in their aesthetic appeal. Of course, some new turbines are placed well offshore and out of sight of people on shore. But in the rural landscapes where they are typically located, very large wind turbines tend to be out of scale with their surroundings. In these common situations, some attempt should be made to balance them with other natural and man-made structures in the landscape.

The principle of site sensitivity involves other considerations than those of scale. It also has to do with the histories and cultural practices of particular regions, with the kinds of materials out of which local buildings are constructed, with the character of available light, and with the flora and fauna (particularly the bird populations) which frequent the area. Again, these kinds of consideration are commonplace among professional architects; they need to become commonplace with wind turbine owners and operators.

2. Equity interest. Local landowners should have an equity interest in the turbines on their property and, where possible, be involved in their maintenance and use. Two points were made in this connection. One is that the permitting of wind turbines takes place at the local level; such permitting is more likely to go through if the turbines are locally owned and operated. The cases of Denmark and the Netherlands are relevant. In Denmark, as in Germany, the majority of turbines are owned by farmers. Easily available low-interest loans and high government-subsidized utility buyback rates make this possible. And in Denmark (the situation in Germany is more complicated), wind energy enjoys very wide acceptance. Indeed, wind power has become identified with the country's high population density, and the Danes are proud of their world leadership in both wind energy production and technology. Much like neighboring Denmark, the Netherlands is a coastal low-lying country with a long windmill tradition and few alternative sources of power. But resistance to wind turbines, largely on aesthetic grounds, is evident in the Netherlands, and it now seems certain that national goals for the use of renewable energy will not be met on schedule. The only evident difference between the two countries is that in the Netherlands wind turbines are for the most part owned by large corporations. There is little local equity.

The other point about local ownership and control is more difficult to describe. It has little to do with the political and economic question. That is, it is not merely that the local people should enjoy the economic benefits of wind energy and therefore want to encourage it. We contend that the system of ownership and control has itself certain aesthetic dimensions. To the extent that someone else, in particular a large corporation, owns an object, it has become alienated from us and we can no longer fully enjoy it. But whatever one's aesthetic theory, even if one has no aesthetic theory at all and thinks that it is all a matter of taste, there is some sort of connection between the perception of beauty and personal enjoyment. Indeed, there is a great deal of evidence that we enjoy and appreciate most that which is near at hand and familiar, those things in which we have a personal stake.

Significant new wind power development will require large sums of money, and the sheer size and complexity of the technology coming on line will require a great deal of engineering expertise. Whether the Danish model of local ownership and control can be widely exported remains to be seen. Short of that, every effort will have to be made to involve local people (in whose hands rests the ultimate fate of wind projects) in every phase of the planning. Such involvement takes patience on all sides and substantial time. However, it is presumably the only way in which people will accept wind energy in their own backyards.

3. Aesthetics. The aesthetic issue needs to be addressed directly. There is, of course, the strategic point that it is an issue best avoided because we believe that beauty *is* in the eye of the beholder. Such a position does little to lessen public resistance to wind energy. This is a subjectivist position to which technocrats and their corporate sponsors would like to retreat. But it is at just this point that we lose people (enough to make a difference) who otherwise support wind energy. They resist because they do not want their own (or personally cherished) area ruined. Urbanites and even major environmental organizations in a variety of different countries are on record against further wind power development. In almost every case, the underlying reasons are aesthetic.

What, then, is the best approach to the aesthetic question? We believe that we cannot deal with it by setting out universal standards of beauty and then compelling designers, developers, or the general public to honor them. Even in our group, highly homogeneous with respect to education, social and economic status, and occupation, there was too much difference of opinion for this to happen. But short of consensus, aesthetic considerations can be made an important and explicit part of the discussion at every level, from the original design of turbines to their eventual installation.

This will involve bringing humanists and artists into the discussions, and breaking down the otherwise crippling distinctions that exist among the arts, humanities, and sciences. We were able to break them down in our own discussions, and this should provide a model for others.

The past few years have witnessed the installation of a great deal of additional wind generating capacity, most of it outside the United States, making wind power the most rapidly growing form of renewable electrical energy in the world. Judging from the shortfalls in electricity supply reported from every quarter of the United States and many other countries, it has been coming at a good time. Even with this boom, however, there is still stubborn resistance to wind energy. For many people, there is no more than grudging acceptance of wind energy's potential. The emphasis, at least in North America, still rests on the notion of finding new sources of fossil fuels rather than on developing renewable energy, an emphasis which the newly elected President of the United States has made clear. This resistance continues to be rooted in aesthetic considerations. People generally are no more ready now than they were when we met in Bellagio to countenance wind turbines erected within their view. Neither are they willing to make radical changes in their lifestyles or reduce their consumption of energy. This is to say that, however large the wind industry's success may appear, we have not yet made a sufficient place for wind turbines in the landscape.

At that, there seemed in our discussions to be a fairly large measure of agreement concerning what was beautiful. Despite all of the arguments about aesthetics, and the frequent allusions to differences in cultural standards and matters of taste, all of us thought that the view from our conference center overlooking Lake Como to the small villages dotting its shores and finally to its mountain enclosure was marvelous. It is a vision, though only one vision, of Arcadia. But there is still the very real problem of fitting wind turbines into any of them, of balancing nature and need. Resolving it will take us deeper into aesthetics, technology, and the landscape. This book has been our contribution to advancing that discussion.

NOTE

1. Where the generator, gearbox, and brakes are housed.

AUTHOR BIOGRAPHIES

EDITORS

Martin J. Pasqualetti is a Professor of Geography at Arizona State University. His primary interest during 30 years of teaching and research has been the complex relationships between energy and land, centering on the territorial requirements of alternative energy resources such as geothermal, solar, and wind. He has also considered long-term warning strategies for nuclear waste repositories, the spatial consequences of nuclear power plant decommissioning, and the environmental costs of energy development along the 2000-mile strip of territory between the United States and Mexico. He has advised such organizations as the Natural Resource Defense Council, Resources for the Future, the U.S. Department of Energy, and the Office of Technology Assessment of the U.S. Congress. He was twice elected chairman of the Energy and Environment Speciality Group of the Association of American Geographers, and was named "Environmental Educator of the Year" by the Association of Energy Engineers. He has published 100 articles and four books on various aspects of energy.

Paul Gipe has worked with wind energy since 1976. His experience with the technology runs the gamut from measuring wind resources to installing residential wind turbines. Gipe is best known for his advocacy of wind energy and for his articles and books on the subject. Gipe has sought to popularize the use of wind energy worldwide. In 1998 the American Wind Energy Association named him as the industry's "person of the year," and in 1998 the World Renewable Energy Congress designated Gipe a "pioneer" in renewable energy. His book *Wind Energy Comes of Age* was selected by the Association of College and Research Libraries, for its list of outstanding academic books in 1995. For eight years he represented the American Wind Energy Association on the West Coast of the United States and was the executive director of the Kern Wind Energy Association.

Robert W. Righter is Research Professor of History at Southern Methodist University, following an extensive teaching and writing career at the University of Wyoming and the University of Texas, El Paso. He has received acclaim for writings on American national parks, including *Crucible for Conservation*; *The Struggle for Grand Teton National Parks*. In recent years his research has turned toward energy issues, particularly wind energy. His recent book *Wind Energy in America, a History* has been a significant contribution to the study of wind energy, past and present.

OTHER CONTRIBUTORS

Gordon G. Brittan, Jr., is Regents Professor of Philosophy and Executive Director of the Wheeler Center at Montana State University in Bozeman. He maintains a long-standing interest in the philosophical bases of the environmental movement. For nearly two decades he has fostered several innovative turbine designs in Montana and California, including a unique "Windjammer" wind turbine which uses natural materials as "sails." A 75-kW wind turbine generates electricity for the local grid and more than enough electricity to supply his Montana ranch.

Karin Hammarlund is a social geographer with Hammarlund A. Konsult in Sweden and a research geographer at the School of Economics and Commercial Law, Göteborg University. For 10 years, she has consulted and written on public acceptance and planning procedures for the promotion of public acceptance of wind energy. She participated in two major wind power investigations for the Swedish National Energy Administration (Vindkraft I harmoni, 1998) and the Environmental Ministry (Vindkraftutredningen, 1999). Hammarlund is presently acting as head of research project, concerning socio-technical aspects of wind power, within the Swedish Wind Energy research Program (VKK).

Martin Hoppe-Kilpper is an electrical engineer specializing in the study of power, control, and measurement engineering. Since 1990 he has been the head of the wind energy department of the Institute for Solar Energy Supply Technology, ISET (Institut für Solare Energieversorgungstechnik), in Kassel, Germany. In this responsibility he is the project manager of the German government's 250-MW Wind Program. He has numerous publications on the present and future prospects of wind energy technology, and he often serves as an external expert for the Research program of the European Commission.

Frode Birk Nielsen is a landscape architect in Aahus, Denmark. His discussion of the architectural and aesthetic characteristics of wind turbines while a student 20 years ago was pioneering. He authored the beautifully illustrated *Wind Turbines and the Landscape: Architecture and Aesthetics*. His firm, Birk Nielsens Tegnestue, has for many years worked on landscaping solutions for the design, visualization and location of wind turbines in the open landscape, in technological landscapes, and in open-sea areas.

Christoph Schwahn is a landscape architect in Göttingen, Germany. His study of the aesthetic impacts of wind turbines in the Weser-Marsh landscape of Lower Saxony was the first of its kind in Germany, and it initiated continuing interest in the growing role which wind turbines play in the landscape in his densely settled and energy progressive country.

Laurence Short brings an artist's perspective to the wind power and landscape debate. Short lived for many years in the famously picturesque Lake District of northwest England, an area so coveted by wind developers that it already hosts 11 wind farm projects, with more planned. Through his firm, the Visual Arts Development Agency, Short bridges the intellectual gaps among art, architecture, and public perceptions of landscape in the United Kingdom.

Urta Steinhäuser (not a participant in Bellagio) is a landscape planner who has worked in diverse city planning offices in Bremen, Cologne, and Melsungen. She has lectured on the subject of monetary compensation for damage done to the environment caused by building construction. Her company "StadtLandFluss" (city/country/river) has contributed to many projects, including the landscaping of several wind parks. This work stimulated a critical dispute about our aptitude to develop environmental preservation standards. She is a member of the architects' guild in Hessen.

BRIEF READING LIST

The amount of reading material is vast and grows every day. What follows, however, is a list of recommended books on the subjects of wind energy, alternative energy, and landscape, not just in the United States but in Europe as well. The footnotes and endnotes in the listed books, as well as in this list, will lead interested readers to a much-expanded bibliography.

Alt, Franz, Jürgen Claus, and Herman Scheer, eds. *Windiger Protest: Konflikte um das Zukunftspotential der Windkraft.* Bochum, Germany: Ponte Press, 1998.

Asmus, Peter. *Reaping the Wind: How Mechanical Wizards, Visionaries, and Profiteers Helped Shape Our Energy Future.* Washington, D.C: Island Press, 2001.

Baker, T. Lindsay. *A Field Guide to American Windmills.* Norman, Oklahoma: University of Oklahoma Press, 1985.

Berleant, Arnold. *The Aesthetics of Environment.* Philadelphia: Temple University Press, 1992.

Borgmann, Albert. *Technology and the Character of Contemporary Life.* Chicago: University of Chicago Press, 1984, reprinted 1987.

Clark, Ronald W. *Works of Man.* New York: Viking Penguin, 1985.

Elliott, David. *Energy; Society and Environment.* London: Routledge, 1997.

Gipe, Paul. *Wind Energy Comes of Age.* New York: John Wiley & Sons, 1995.

Francaviglia, Richard. *Hard Places.* Iowa City: University of Iowa Press, 1991.

Hirsh, Richard F. *Technology and Transformation in the American Electric Utility Industry.* New York: Cambridge University Press, 1989.

Hoskins, W.G. *The Making of the English Landscape.* London: Penguin, 1999.

Hussey, Christopher. *The Picturesque: Studies in a Point of View.* New York G.P. Putnam's Sons, 1927.

Jackson, John Brinckerhoff. *A Sense of Place, a Sense of Time.* New Haven: Yale University Press, 1994.

_____ *Discovering the Vernacular Architecture*. New Haven: Yale University Press, 1984.

Kealey, Edward J. *Harvesting the Air: Windmill Pioneers in Twelfth-Century England*. Berkeley: University of California Press, 1987.

Leopold, Aldo. *A Sand County Almanac and Sketches Here and There*. Oxford: Oxford University Press, 1989.

_____ *Round River, From the Journals of Aldo Leopold*. Oxford: Oxford University Press, 1972.

Lovins, Amory. *Soft Energy Paths: Toward a Durable Peace*. Cambridge, MA: Ballinger, 1977.

Marx, Leo. *The Machine in the Garden*. New York: Oxford University Press, 1964.

McHarg, Ian. *Design with Nature*. New York: John Wiley & Sons, 1995 ed.

Nielsen, Frode Birk. *Wind Turbines & the Landscape: Architecture & Aesthetics*. Århus, Denmark: Birk Nielsens Tegnestue, 1996.

Porteous, J. Douglas. *Environmental Aesthetics: Ideas, Politics and Planning London*. New York: Routledge, 1996.

Schama, Simon. *Landscape and Memory*. New York: Vintage Books, 1996.

Righter, Robert W. *Wind Energy in America*. Norman, Oklahoma: University of Oklahoma Press, 1996.

Roe, David. *Dynamos and Virgins*. New York: Random House, 1984.

Schumacher, E. F. *Small Is Beautiful: Economics as If People Mattered*. New York: Harper & Row, 1973.

Shepard, Eugene S. *Engineering and the Mind's Eye*. Cambridge, MA: MIT Press, 1992.

Shepard, Paul. *Man in the Landscape*. New York: Alfred Knopf, 1967.

Stanton, Caroline. *The Landscape Impact and Visual Design of Windfarms*, School of Landscape Architecture, Edinburgh College of Art, Heriot-Watt University, Lauriston Place; Edinburgh EH3 9DF, Scotland, United Kingdom, 1996.

Thayer, Robert Jr., *Gray World, Green Heart*. New York: John Wiley, 1994.

Torrey, Volta. *Wind-Catchers: American Windmills of Yesterday and Today*. Brattleboro, VT: Stephen Greene, 1976.

Tuan, Yi-Fu. *Topophilia: A Study of Environmental Perceptions, Attitudes, and Values*. New York: Columbia University Press, 1974, 1990 ed.

Urry, John. *Consuming Places*. London: Routledge, 1995.

INDEX